I0390386

ISBN 978-1-291-43617-4

In copertina foto cometa West ©
On the cover photo comet West

1

INTRODUZIONE

Questo libro, il decimo di una serie di dieci, rappresenta una estesa trattazione di quanto presente sul mio sito riguardo le comete ed i fenomeni ad esse correlate. Vengono qui esaminati gli elementi orbitali delle comete periodiche e non periodiche, con le date dei perielii per decenni, le magnitudini, l'evoluzione degli elementi, le classiche effemeridi, ecc.

Questo non è un manuale tecnico e di difficile lettura, ma una descrizione completa e molto dettagliata su quello che il cielo ci offre durante la nostra vita, quindi ogni tabella è pronta all'uso ed ogni evento riportato sarà facilmente visibile ad occhio nudo od eventualmente con un modestissimo binocolo.

Un'opera per astrofili, per astronomi, per professionisti o semplici appassionati.

4

INTRODUCTION

This book, the tenth in a series of ten, is an extended discussion of that on my website about the comets. There are all the aspects of the orbital elements of the periodical and not-periodical comets, with the dates of the perihelium for many decades, the magnitudes, the evolution of the elements, with ephemerides, ecc.
This is not a technical and difficult to read manual, but a complete and very detailed description of what the sky gives us throughout our lives, so each table is ready for use, and each reported event will be easily visible to the naked eye or possibly with a simple pair of binoculars.
The book is for stargazing astronomers and professionals.

ELEMENTI ORBITALI
ORBITAL ELEMENTS

Perihelium : date in the format yyyy/mm/dd
q : distance in A.U.
e : eccentricity
w : argument of the perihelum
Ω : longitude of the node
I : inclination
G   H : magnitude parameters

Perihelium : data del perielio nel formato aaaa/mm/gg
q : distanza dal Sole in Unità Astronomiche
e : eccentricità
w : argument del perielio
Ω : longitudine del nodo ascendente
I : inclinazione
G   H : parametric della magnitudine

Epoca — epoch : J2000

Name (nome)	Perihelium	q	e
1P/Halley	1986 02 05.9000	0.586000	0.967000
2P/Encke	2013 11 21.6940	0.336127	0.848232
3D/Biela	1832 11 26.6200	0.879000	0.751000
4P/Faye	2014 05 29.5253	1.655402	0.568711
5D/Brorsen	1879 03 31.0300	0.590000	0.810000
6P/d'Arrest	2008 08 15.0259	1.353307	0.612704
7P/Pons-Winnecke	2008 09 26.6251	1.253563	0.634864
8P/Tuttle	2008 01 27.0243	1.026929	0.819754
9P/Tempel	2011 01 12.4399	1.512570	0.516333
10P/Tempel	2015 11 15.1934	1.420985	0.536363
11P/Tempel-Swift-LINEAR	2008 05 04.5768	1.553565	0.544949
12P/Pons-Brooks	1954 05 22.8800	0.774000	0.955000
13P/Olbers	1956 06 19.1400	1.178000	0.930000
14P/Wolf	2009 02 26.6844	2.723311	0.359017
15P/Finlay	2008 06 22.4872	0.968176	0.721732
16P/Brooks	2014 06 07.6533	1.466222	0.562958
17P/Holmes	2014 03 27.2555	2.055545	0.432387
18D/Perrine-Mrkos	1968 11 01.5400	1.272000	0.643000
19P/Borrelly	2008 07 22.5475	1.356214	0.624181
20D/Westphal	1913 11 26.7900	1.254000	0.920000
21P/Giacobini-Zinner	2012 02 11.7324	1.030484	0.707054
22P/Kopff	2009 05 25.3762	1.578029	0.544257
23P/Brorsen-Metcalf	1989 09 11.9400	0.479000	0.972000
24P/Schaumasse	2009 08 9.6721	1.213719	0.703483
25D/Neujmin 2	1927 01 16.2241	1.338169	0.566818
26P/Grigg-Skjellerup	2013 07 05.9954	1.086353	0.640143
27P/Crommelin	2011 08 03.8171	0.747948	0.918822
28P/Neujmin	2002 12 26.7785	1.556918	0.774882
29P/Schwassmann-Wachmann	2019 04 08.3700	5.734351	0.043971
30P/Reinmuth	2010 04 19.6134	1.881970	0.501413
31P/Schwassmann-Wachmann	2010 09 29.7124	3.423909	0.193565
32P/Comas Sola	2014 10 17.7240	2.003345	0.555637
33P/Daniel	2008 07 20.3005	2.169362	0.461966
34D/Gale	1938 06 18.4832	1.182907	0.760719
35P/Herschel-Rigollet	1939 08 09.4640	0.748490	0.974050
36P/Whipple	2011 12 29.2521	3.087774	0.261484
37P/Forbes	2011 12 11.0267	1.575315	0.540664
38P/Stephan-Oterma	1980 12 05.1761	1.574420	0.860022
39P/Oterma	2002 12 21.0203	5.471147	0.245513
40P/Vaisala 1	2004 01 22.8860	1.795974	0.632912
41P/Tuttle-Giacobini-Kresak	2011 11 12.1468	1.049404	0.660196
42P/Neujmin 3	2004 07 15.9444	2.014714	0.585156
43P/Wolf-Harrington	2010 07 01.9327	1.357234	0.594830
44P/Reinmuth	2015 03 24.1545	2.116805	0.426820
45P/Honda-Mrkos-Pajdusakova	2011 09 28.7827	0.529704	0.824658
46P/Wirtanen	2013 07 09.4225	1.052753	0.659174
47P/Ashbrook-Jackson	2017 06 10.5688	2.810571	0.317545
48P/Johnson	2011 09 29.3987	2.301226	0.367917
49P/Arend-Rigaux	2011 10 19.0858	1.423884	0.600364
50P/Arend	2007 11 1.1975	1.924341	0.529093
51P/Harrington	2008 06 19.4163	1.687664	0.544478
52P/Harrington-Abell	2014 03 07.6745	1.773290	0.540248
53P/Van Biesbroeck	2003 10 09.4450	2.414868	0.552290
54P/de Vico-Swift-NEAT	2009 11 28.5134	2.171755	0.427030
55P/Tempel-Tuttle	1998 02 28.0977	0.976427	0.905552
56P/Slaughter-Burnham	2005 01 14.9297	2.534969	0.503671
57P/du Toit-Neujmin-Delporte	2008 12 26.0489	1.723717	0.500077
58P/Jackson-Neujmin	2012 04 10.0133	1.374439	0.662573
59P/Kearns-Kwee	2009 03 7.8897	2.355957	0.474519
60P/Tsuchinshan	2012 05 13.5172	1.618168	0.538514
61P/Shajn-Schaldach	2015 10 01.8529	2.112018	0.426349
62P/Tsuchinshan	2011 06 30.4153	1.383587	0.597594

Name (nome)	Perihelium	q	e
63P/Wild	2013 04 10.7060	1.950774	0.650892
64P/Swift-Gehrels	2009 06 14.295	1.3770	0.689544
65P/Gunn	2010 03 09.0233	2.458436	0.321476
66P/du Toit	2003 08 27.9919	1.274266	0.787701
67P/Churyumov-Gerasimenko	2009 02 28.9290	1.263790	0.637770
68P/Klemola	2009 01 21.2612	1.758125	0.640212
69P/Taylor	2011 07 17.5688	2.272907	0.414468
70P/Kojima	2007 10 5.9505	2.011841	0.453285
71P/Clark	2011 12 15.9046	1.567473	0.498445
72D/Denning-Fujikawa	1978 10 02.1422	0.779727	0.819859
73P/Schwassmann-Wachmann	2011 10 16.8988	0.942870	0.692248
74P/Smirnova-Chernykh	2009 07 26.9842	3.552746	0.147426
75D/Kohoutek	1994 06 29.8984	1.784656	0.496307
76P/West-Kohoutek-Ikemura	2013 05 07.7686	1.600269	0.538849
77P/Longmore	2016 05 14.8157	2.329935	0.354672
78P/Gehrels	2012 01 12.8628	2.008540	0.462750
79P/du Toit-Hartley	2013 08 23.2709	1.123942	0.618599
80P/Peters-Hartley	2006 09 25.7902	1.633571	0.596013
81P/Wild	2010 02 22.4793	1.597589	0.537086
82P/Gehrels	2010 01 13.5258	3.633049	0.123311
83D/Russell 1	1985 07 05.2291	1.611541	0.517209
84P/Giclas	2013 07 23.1161	1.840173	0.494468
85P/Boethin	2008 12 16.3259	1.147535	0.775555
86P/Wild	2008 05 20.0346	2.301179	0.365687
87P/Bus	2013 12 19.3018	2.103157	0.388893
88P/Howell	2015 04 06.8906	1.361980	0.562058
89P/Russell	2009 08 17.1356	2.280621	0.399055
90P/Gehrels 1	2002 06 23.1147	2.965911	0.509136
91P/Russell	2013 03 01.4082	2.616388	0.328834
92P/Sanguin	2002 09 22.9975	1.808040	0.663082
93P/Lovas	2007 12 17.3624	1.704754	0.611620
94P/Russell	2010 03 29.8580	2.239971	0.363010
95P/Chiron	1996 02 4.2835	6.939300	0.380160
96P/Machholz	2012 07 14.7859	0.123792	0.959181
97P/Metcalf-Brewington	2011 08 20.8775	2.596662	0.458751
98P/Takamizawa	2013 08 05.4088	1.673418	0.560589
99P/Kowal	2007 01 16.0982	4.718358	0.227534
100P/Hartley	2009 12 6.1405	1.982377	0.418754
101P/Chernykh	2005 12 25.3388	2.350622	0.594171
102P/Shoemaker	2013 08 31.9244	1.968536	0.473177
103P/Hartley 2	2010 10 28.2781	1.058690	0.695128
104P/Kowal	2010 05 04.5040	1.178977	0.638756
105P/Singer Brewster	2012 02 26.1574	2.050860	0.409270
106P/Schuster	2007 04 2.2217	1.556134	0.586802
107P/Wilson-Harrington	2009 10 22.0544	0.991729	0.624269
108P/Ciffreo	2007 07 18.0528	1.719104	0.541486
109P/Swift-Tuttle	1992 12 11.9997	0.959516	0.963225
110P/Hartley	2014 12 17.6073	2.478124	0.313976
111P/Helin-Roman-Crockett	2013 02 07.5224	3.698262	0.115424
112P/Urata-Niijima	2013 06 24.3122	1.455331	0.588044
113P/Spitaler	2008 03 23.4814	2.128435	0.422807
114P/Wiseman-Skiff	2013 05 13.9355	1.574871	0.555400
115P/Maury	2011 10 06.9687	2.035131	0.520948
116P/Wild	2016 01 12.7225	2.182322	0.372824
117P/Helin-Roman-Alu	2014 03 28.0900	3.053588	0.253847
118P/Shoemaker-Levy	2010 01 3.5720	1.995203	0.425207
119P/Parker-Hartley	2014 04 02.1361	3.028759	0.292059
120P/Mueller	2013 02 22.3640	2.729357	0.338598
121P/Shoemaker-Holt	2013 09 16.4263	3.749449	0.190634
122P/de Vico	1995 10 06.0001	0.659337	0.962708
123P/West-Hartley	2011 07 04.4830	2.128786	0.448584
124P/Mrkos	2014 04 09.6121	1.643586	0.504227

Name (nome)	Perihelium	q	e
125P/Spacewatch	2013 02 17.0266	1.525470	0.512150
126P/IRAS	2010 02 22.8393	1.713300	0.696401
127P/Holt-Olmstead	2009 10 21.3258	2.195724	0.362704
128P/Shoemaker-Holt	1988 05 21.3664	3.052969	0.321852
129P/Shoemaker-Levy	2014 01 21.4191	3.886746	0.108212
130P/McNaught-Hughes	2011 06 24.9977	2.098291	0.407357
131P/Mueller	2012 01 07.2382	2.418005	0.343929
132P/Helin-Roman-Alu 2	2006 02 14.9808	1.924157	0.530008
133P/Elst-Pizarro	2013 02 09.0792	2.649736	0.161803
134P/Kowal-Vavrova	2014 05 21.3952	2.572923	0.587028
135P/Shoemaker-Levy	2014 11 01.2088	2.683573	0.294515
136P/Mueller	2016 05 29.5979	2.968886	0.292714
137P/Shoemaker-Levy	2009 05 12.7593	1.889263	0.578431
138P/Shoemaker-Levy	2012 06 11.7660	1.700663	0.530478
139P/Vaisala-Oterma	2017 12 09.2650	3.406106	0.247451
140P/Bowell-Skiff	1999 05 14.8119	1.971877	0.691761
141P/Machholz	2010 05 29.7154	0.757705	0.749138
142P/Ge-Wang	2010 05 29.8214	2.487652	0.498739
143P/Kowal-Mrkos	2009 06 11.5446	2.542784	0.408627
144P/Kushida	2009 01 26.8933	1.438989	0.627670
145P/Shoemaker-Levy	2009 03 26.4253	1.892029	0.542607
146P/Shoemaker-LINEAR	2008 05 21.3753	1.417460	0.647993
147P/Kushida-Muramatsu	2008 09 23.1507	2.756207	0.275664
148P/Anderson-LINEAR	2008 05 22.7979	1.702390	0.537767
149P/Mueller	2010 02 19.2808	2.650716	0.388630
150P/LONEOS	2008 11 25.9636	1.766533	0.546112
151P/Helin	2001 09 23.3590	2.531841	0.566295
152P/Helin-Lawrence	2012 07 09.2910	3.116417	0.307416
153P/Ikeya-Zhang	2002 03 18.9807	0.507141	0.990097
154P/Brewington	2013 12 12.1988	1.608044	0.670648
155P/Shoemaker 3	2002 12 14.8294	1.813342	0.726672
156P/Russell-LINEAR	2014 04 16.3747	1.585599	0.559204
157P/Tritton	2016 06 09.8414	1.357753	0.601881
158P/Kowal-LINEAR	2012 09 22.9691	4.576599	0.030449
159P/LONEOS	2018 05 21.7395	3.619025	0.382557
160P/LINEAR	2012 09 18.4866	2.066624	0.479115
161P/Hartley-IRAS	2005 06 20.9471	1.272193	0.835102
162P/Siding Spring	2015 07 11.4947	1.235382	0.595697
163P/NEAT	2012 04 12.7500	2.056678	0.453464
164P/Christensen	2011 06 02.3633	1.675617	0.541283
165P/LINEAR	2000 06 15.7921	6.830371	0.621584
166P/NEAT	2002 05 24.2686	8.571168	0.382741
167P/CINEOS	2001 05 09.7229	11.837502	0.271397
168P/Hergenrother	2012 10 01.6760	1.415042	0.609392
169P/NEAT	2014 02 15.0702	0.607608	0.766925
170P/Christensen	2006 01 26.5843	2.928445	0.304170
171P/Spahr	2012 04 30.5203	1.764633	0.503326
172P/Yeung	2008 10 12.6011	2.240903	0.362098
173P/Mueller	2008 05 15.7010	4.203542	0.261390
174P/Echeclus	2015 04 22.2624	5.815664	0.457106
175P/Hergenrother	2013 05 23.5716	1.946331	0.432138
176P/LINEAR	2011 06 29.8687	2.575532	0.193893
177P/Barnard	2006 08 28.6860	1.107242	0.954404
178P/Hug-Bell	2013 07 22.9957	1.934027	0.473025
179P/Jedicke	2007 12 3.2203	4.086730	0.307562
180P/NEAT	2008 05 26.5968	2.468677	0.357980
181P/Shoemaker-Levy	2014 06 10.0286	1.122634	0.707725
182P/LONEOS	2012 03 05.4458	1.008530	0.659456
183P/Korlevic-Juric	2017 11 10.1625	3.884216	0.135407
184P/Lovas	2013 07 28.4658	1.394032	0.604345
185P/Petriew	2012 08 13.5468	0.931888	0.699431
186P/Garradd	2008 05 27.9657	4.328669	0.125925

Name (nome)	Perihelium	q	e
187P/LINEAR	2008 10 28.1513	3.752293	0.169509
188P/LINEAR-Mueller	2007 12 16.1477	2.552193	0.415686
189P/NEAT	2012 07 20.4283	1.177232	0.596783
190P/Mueller	2007 07 08.2051	2.031918	0.520767
191P/McNaught	2014 05 06.1339	2.044243	0.420920
192P/Shoemaker-Levy	2007 12 17.3351	1.460073	0.773603
193P/LINEAR-NEAT	2008 02 20.4642	2.155772	0.396097
194P/LINEAR	2008 02 26.1850	1.708744	0.574149
195P/Hill	2009 01 20.3116	4.433857	0.315378
196P/Tichy	2008 02 07.1550	2.137865	0.433899
197P/LINEAR	2013 03 24.8998	1.061639	0.629583
198P/ODAS	2012 02 15.7802	1.996592	0.444825
199P/Shoemaker	2009 04 9.9709	2.935397	0.507141
200P/Larsen	2008 08 26.2764	3.280490	0.333109
201P/LONEOS	2008 08 4.7584	1.345001	0.611677
202P/Scotti	2009 02 7.0219	2.527052	0.330740
203P/Korlevic	2010 02 09.1406	3.183231	0.316072
204P/LINEAR-NEAT	2008 12 9.2711	1.940235	0.470561
205P/Giacobini	2008 09 10.0558	1.526442	0.568719
206P/Barnard-Boattini	2008 10 25.0956	1.145196	0.646480
207P/NEAT	2008 11 6.2547	0.944150	0.757147
208P/McMillan	2008 05 13.3110	2.525052	0.374412
209P/LINEAR	2009 04 15.9729	0.913703	0.688980
210P/Christensen	2008 12 19.9718	0.534907	0.831644
211P/Hill	2009 05 7.8248	2.362033	0.337560
212P/NEAT	2008 12 3.2697	1.654465	0.578876
213P/Van Ness	2011 06 16.4833	2.122742	0.380247
214P/LINEAR	2009 01 5.6146	1.843785	0.488611
215P/NEAT	2010 06 17.5079	3.222209	0.205489
216P/LINEAR	2008 10 10.9207	2.159822	0.443938
217P/LINEAR	2009 09 8.9657	1.223968	0.689679
218P/LINEAR	2009 06 22.2864	1.701439	0.490270
219P/LINEAR	2010 03 04.9058	2.363899	0.352330
220P/McNaught	2009 12 15.4652	1.548585	0.502540
221P/LINEAR	2009 01 24.8669	1.783572	0.487491
222P/LINEAR	2009 09 01.1051	0.780162	0.727021
223P/Skiff	2010 08 14.5513	2.421376	0.416336
224P/LINEAR-NEAT	2010 01 31.7122	1.989859	0.416679
225P/LINEAR	2009 08 25.5274	1.314741	0.639325
226P/Pigott-LINEAR-Kowalski	2009 05 11.2515	1.769207	0.529986
227P/Catalina-LINEAR	2010 09 03.7223	1.794826	0.499869
228P/LINEAR	2011 08 23.9416	3.430476	0.177562
229P/Gibbs	2009 08 04.2214	2.440269	0.377926
230P/LINEAR	2009 08 08.8333	1.485982	0.562902
231P/LINEAR-NEAT	2011 05 16.4627	3.032488	0.246785
232P/Hill	2009 10 01.4918	2.983174	0.334668
233P/La Sagra	2010 03 12.4093	1.791436	0.409562
234P/LINEAR	2009 12 23.3769	2.861043	0.250802
235P/LINEAR	2010 03 21.7827	2.748252	0.313613
236P/LINEAR	2010 09 08.8328	1.831338	0.509083
237P/LINEAR	2009 12 06.0204	2.386903	0.359567
238P/Read	2011 03 10.0764	2.359881	0.253660
239P/LINEAR	2000 02 17.0532	1.645062	0.631525
240P/NEAT	2010 10 04.4774	2.124721	0.450202
241P/LINEAR	2010 07 18.3009	1.921410	0.610975
242P/Spahr	2012 04 03.5456	3.979926	0.279123
243P/NEAT	2011 03 02.7210	2.455193	0.359128
244P/Scotti	2012 01 20.4255	3.918208	0.199781
245P/WISE	2010 02 04.6438	2.139607	0.466517
246P/NEAT	2013 01 28.9622	2.879374	0.285035
247P/LINEAR	2011 01 04.1445	1.484393	0.625478
248P/Gibbs	2011 02 08.8062	2.147574	0.640446

Name (nome)	Perihelium	q	e
249P/LINEAR	2011 04 16.0809	0.510880	0.816033
250P/Larson	2010 11 14.3720	2.214037	0.406669
251P/LINEAR	2010 12 29.5242	1.714283	0.509026
252P/LINEAR	2000 03 09.9580	1.001368	0.672411
253P/PANSTARRS	2011 11 24.0413	2.039334	0.412863
254P/McNaught	2010 10 25.5918	3.214070	0.312905
255P/Levy	2012 01 14.9440	1.007477	0.668271
256P/LINEAR	2013 03 17.3639	2.689922	0.418785
257P/Catalina	2013 06 04.4310	2.129041	0.432778
258P/PANSTARRS	2011 03 19.1199	3.479202	0.208804
259P/Garradd	2013 01 25.4258	1.797632	0.341109
260P/McNaught	2012 09 12.5155	1.497154	0.593459
261P/Larson	2012 09 29.0883	2.186967	0.389574
262P/McNaught-Russell	2012 12 04.4771	1.279998	0.815338
263P/Gibbs	2006 12 28.7660	1.251302	0.586980
264P/Larsen	2011 11 21.8437	2.449856	0.372363
265P/LINEAR	2012 06 10.7563	1.498918	0.647057
266P/Christensen	2013 09 01.0275	2.327221	0.340794
267P/LONEOS	2012 08 22.7997	1.336179	0.593515
268P/Bernardi	2005 08 12.6846	2.345352	0.478874
269P/Jedicke	2014 11 16.9764	4.077839	0.438303
270P/Gehrels	2013 07 08.9166	3.601112	0.472540
271P/van Houten-Lemmon	2013 07 05.8414	4.250046	0.390703
272P/NEAT	2013 02 27.1566	2.416712	0.455860
273P/Pons-Gambart	2012 12 19.6590	0.810196	0.975320
274P/Tombaugh-Tenagra	2013 02 23.3430	2.441867	0.440307
275P/Hermann	2012 12 27.2812	1.643765	0.714136
276P/Vorobjov	2013 01 15.9305	3.923838	0.273319
277P/LINEAR	2013 06 05.9370	1.913177	0.504416
278P/McNaught	2013 08 02.5295	2.097660	0.433164
279P/La Sagra	2009 10 10.5166	2.147705	0.399352
280P/Larsen	2013 12 11.1172	2.635885	0.417229
281P/MOSS	2012 05 15.0109	4.014355	0.173030
C/1995 O1 (Hale-Bopp)	1997 03 30.4718	0.926799	0.994981
P/1998 QP54 (LONEOS-Tucker)	2007 05 12.1870	1.879916	0.552000
P/1998 U4 (Spahr)	2012 04 29.0477	3.860593	0.308588
P/1998 VS24 (LINEAR)	2008 05 24.9893	3.422409	0.241676
P/1999 XN120 (Catalina)	2008 11 13.4526	3.302978	0.210406
P/2000 Y3 (Scotti)	2011 12 30.4643	4.018632	0.196925
C/2001 HT50 (LINEAR-NEAT)	2003 07 9.1615	2.785122	0.997781
C/2001 K5 (LINEAR)	2002 10 13.0561	5.188534	1.001280
C/2001 Q4 (NEAT)	2004 05 16.0218	0.964376	1.000934
C/2002 J5 (LINEAR)	2003 09 20.1016	5.735457	1.000027
C/2002 L9 (NEAT)	2004 04 6.2318	7.033192	0.999013
C/2002 T7 (LINEAR)	2004 04 22.5183	0.617486	1.000449
C/2002 VQ94 (LINEAR)	2006 02 6.9704	6.797191	0.965722
C/2003 A2 (Gleason)	2003 11 9.877811	431815	1.002631
C/2003 E1 (NEAT)	2004 02 14.1174	3.248731	0.764609
C/2003 K4 (LINEAR)	2004 10 13.8974	1.023973	1.000638
C/2003 T3 (Tabur)	2004 04 28.7094	1.479955	0.999146
C/2003 T4 (LINEAR)	2005 04 3.6505	0.850607	1.000465
C/2003 WT42 (LINEAR)	2006 04 10.8547	5.191254	1.001776
P/2004 A1 (LONEOS)	2004 09 6.8126	5.478457	0.312004
C/2004 B1 (LINEAR)	2006 02 7.8340	1.601853	1.001011
C/2004 D1 (NEAT)	2006 02 10.9015	4.975120	1.000656
P/2004 DO29 (Spacewatch-LINEAR	2004 10 4.8577	4.092761	0.443316
P/2004 FY140 (LINEAR)	2004 08 5.5634	4.107600	0.167680
C/2004 K1 (Catalina)	2005 07 5.1062	3.399921	0.998448
C/2004 L2 (LINEAR)	2005 11 15.1788	3.778560	0.995712
C/2004 Q2 (Machholz)	2005 01 24.3201	1.204706	0.998578
C/2004 T3 (Siding Spring)	2003 04 16.7373	8.864851	1.001055
P/2004 V5 (LINEAR-Hill)	2005 03 3.5632	4.418753	0.446310

12

Name (nome)	Perihelium	q	e
P/2004 VR8 (LONEOS)	2005 09 5.6442	2.376531	0.514777
C/2005 B1 (Christensen)	2006 02 23.6070	3.204888	1.000742
C/2005 E2 (McNaught)	2006 02 23.4689	1.519137	1.000067
C/2005 EL173 (LONEOS)	2007 03 5.8684	3.886331	1.003403
C/2005 G1 (LINEAR)	2006 02 27.5846	4.960913	1.000388
P/2005 JD108 (Catalina-NEAT)	2005 08 10.9188	4.029088	0.375330
P/2005 JY126 (Catalina)	2006 02 21.3820	2.125878	0.433691
C/2005 K1 (Skiff)	2005 11 21.2189	3.693525	1.002682
P/2005 L1 (McNaught)	2005 12 13.4933	3.144023	0.209122
C/2005 L3 (McNaught)	2008 01 15.0415	5.594578	1.001279
C/2005 O1 (NEAT)	2005 05 17.3488	3.595351	0.928904
C/2005 O2 (Christensen)	2005 09 8.3325	3.334296	0.858824
C/2005 Q1 (LINEAR)	2005 08 25.1026	6.407817	1.003341
C/2005 R4 (LINEAR)	2006 03 7.7321	5.187905	0.997303
P/2005 RV25 (LONEOS-Christense	2006 11 7.1580	3.607840	0.166242
P/2005 S2 (Skiff)	2006 06 27.9587	6.398145	0.196966
C/2005 S4 (McNaught)	2007 07 18.5735	5.850232	0.998961
P/2005 SB216 (LONEOS)	2007 02 11.3487	3.818424	0.463311
C/2005 W2 (Christensen)	2006 03 27.3340	3.331218	0.823980
P/2005 Y2 (McNaught)	2004 12 28.4880	3.355756	0.467148
C/2005 YW (LINEAR)	2006 12 7.8600	1.993071	0.989638
C/2006 A1 (Pojmanski)	2006 02 22.1738	0.555737	0.999798
C/2006 CK10 (Catalina)	2006 07 3.2691	1.751985	0.991947
C/2006 E1 (McNaught)	2007 01 6.7778	6.040311	1.001602
P/2006 F1 (Kowalski)	2008 02 16.5353	4.119952	0.122964
C/2006 F2 (Christensen)	2006 03 31.5125	4.297983	0.655096
P/2006 G1 (McNaught)	2006 08 18.3764	2.632088	0.453171
P/2006 HR30 (Siding Spring)	2007 01 2.2804	1.226432	0.843122
C/2006 HW51 (Siding Spring)	2006 09 29.2429	2.265426	1.001981
C/2006 K1 (McNaught)	2007 07 20.5855	4.425477	1.001196
C/2006 K3 (McNaught)	2007 03 13.3648	2.501427	1.000870
C/2006 K4 (NEAT)	2007 11 29.3027	3.188663	0.998197
C/2006 L1 (Garradd)	2006 10 18.0016	1.462032	0.997353
C/2006 L2 (McNaught)	2006 11 20.2044	1.994003	1.000869
C/2006 M1 (LINEAR)	2007 02 13.9184	3.556287	0.976972
C/2006 M4 (SWAN)	2006 09 28.7252	0.783015	1.000147
C/2006 OF2 (Broughton)	2008 09 15.7644	2.431806	1.000343
C/2006 P1 (McNaught)	2007 01 12.7982	0.170747	1.000015
C/2006 Q1 (McNaught)	2008 07 3.7776	2.763970	1.000122
P/2006 R2 (Christensen)	2006 06 15.2692	3.039238	0.270992
C/2006 S2 (LINEAR)	2007 05 7.3624	3.161484	1.002125
C/2006 S3 (LONEOS)	2012 04 14.0537	5.139239	1.001624
P/2006 S4 (Christensen)	2006 06 1.7259	3.068237	0.508294
C/2006 S5 (Hill)	2007 12 9.7733	2.629527	0.973328
P/2006 T1 (Levy)	2006 10 7.4315	0.989595	0.672105
P/2006 U5 (Christensen)	2007 01 18.9043	2.325904	0.341035
C/2006 U6 (Spacewatch)	2008 06 5.4259	2.498045	0.998687
C/2006 V1 (Catalina)	2007 11 26.5307	2.675090	0.989495
C/2006 VZ13 (LINEAR)	2007 08 10.8916	1.015259	1.000238
C/2006 W3 (Christensen)	2009 07 6.2166	3.126690	1.000778
C/2006 WD4 (Lemmon)	2007 04 28.3983	0.591238	0.998983
C/2006 X1 (LINEAR)	2006 03 5.8045	6.126164	1.000103
C/2006 XA1 (LINEAR)	2007 07 21.8923	1.804292	0.992822
P/2006 XG16 (Spacewatch)	2007 02 9.9500	2.102211	0.420945
C/2006 YC (Catalina-Christense	2006 09 11.6988	4.948063	1.000210
P/2007 B1 (Christensen)	2007 01 19.9563	2.442775	0.580935
C/2007 B2 (Skiff)	2008 08 20.4699	2.974942	0.996995
P/2007 C1 (Christensen)	2007 03 5.2603	2.050793	0.412494
P/2007 C2 (Catalina)	2007 09 4.7068	3.779640	0.462240
C/2007 D1 (LINEAR)	2007 06 21.1345	8.792481	1.002299
C/2007 D3 (LINEAR)	2007 05 27.7322	5.209207	0.992103
C/2007 E1 (Garradd)	2007 05 23.8291	1.285868	0.980219

13

Name (nome)	Perihelium	q	e
C/2007 E2 (Lovejoy)	2007 03 27.5124	1.093001	0.999709
C/2007 F1 (LONEOS)	2007 10 28.6476	0.401622	1.000130
C/2007 G1 (LINEAR)	2008 11 16.5513	2.645138	1.000769
P/2007 H1 (McNaught)	2007 08 17.2345	2.282314	0.379796
P/2007 H3 (Garradd)	2007 08 15.3446	1.829275	0.477358
C/2007 JA21 (LINEAR)	2006 11 13.6058	5.365099	1.000000
C/2007 K1 (Lemmon)	2007 04 28.898	9.25276	1.00000
P/2007 K2 (Gibbs)	2007 06 8.548	2.27165	0.69271
C/2007 K3 (Siding Spring)	2008 04 21.831	2.05093	1.00000
C/2007 K4 (Gibbs)	2007 05 3.878	3.52685	1.00000
C/2007 K5 (Lovejoy)	2007 05 1.951	1.14909	0.97654
C/2007 K6 (McNaught)	2007 07 4.647	3.43751	1.00000
C/2007 M1 (McNaught)	2008 08 21.327	7.44479	1.00000
C/2007 M2 (Catalina)	2008 12 8.227	3.51239	1.00000
C/2007 M3 (LINEAR)	2007 09 4.375	3.47310	1.00000
C/2007 N1 (McNaught)	2007 09 7.236	2.28009	1.00000
C/2007 N3 Lulin	2009 01 10.6577	1.209265	0.999581
C/2007 O1 LINEAR	2007 06 03.2455	2.876674	1.004348
C/2007 P1 McNaught	2007 04 03.862	0.51424	1.00000
C/2007 Q1 Garradd	2006 12 10.844	2.97763	1.00000
P/2007 Q2 Gilmore	2007 08 23.8862	1.839028	0.671227
C/2007 Q3 Siding Spring	2009 10 07.7882	2.251535	0.999149
P/2007 R1 Larson	2007 08 23.1035	4.355628	0.276400
P/2007 R2 Gibbs	2007 08 26.6699	1.465892	0.573963
P/2007 R3 Gibbs	2007 07 06.0410	2.503612	0.416001
P/2007 R4 Garradd	2007 09 27.2798	1.921408	0.671466
P/2007 S1 Zhao	2007 12 06.7033	2.494380	0.343312
C/2007 S2 Lemmon	2008 09 22.9241	5.538857	0.552836
C/2007 T1 McNaught	2007 12 12.519	0.96885	1.00000
P/2007 T2 Kowalski	2007 09 19.028	0.69593	0.77493
P/2007 T4 Gibbs	2007 07 19.460	1.99968	0.60695
C/2007 T5 Gibbs	2008 05 24.683	4.04807	0.91574
P/2007 T6 Catalina	2007 08 20.100	2.23777	0.51939
C/2007 U1 LINEAR	2008 08 07.074	3.33847	1.00000
P/2007 V1 Larson	2007 12 08.4739	2.676584	0.461255
P/2007 V2 Hill	2007 07 30.9904	2.774777	0.318024
C/2007 VO53 Spacewatch	2010 04 26.6855	4.845087	0.999142
P/2007 VQ11 Catalina	2008 02 13.6152	2.693631	0.502150
C/2007 W1 Boattini	2008 06 24.8880	0.849650	1.000137
C/2007 W3 LINEAR	2008 06 02.8155	1.776179	0.999887
C/2007 Y1 LINEAR	2008 03 19.3127	3.340791	1.000336
C/2007 Y2 McNaught	2008 04 08.4077	4.208901	0.996311
C/2008 A1 McNaught	2008 09 29.1343	1.073045	1.000267
P/2008 A2 LINEAR	2008 06 12.0362	1.305407	0.591306
C/2008 C1 Chen-Gao	2008 04 16.8516	1.262338	1.000000
C/2008 E1 Catalina	2008 08 10.9780	4.830160	0.547372
C/2008 E3 Garradd	2008 08 30.828	5.46234	1.00000
C/2008 G1 Gibbs	2009 01 11.9524	3.990335	0.989687
C/2008 H1 LINEAR	2008 03 16.566	2.76108	0.94815
C/2008 J1 Boattini	2008 07 13.387	1.72402	1.00000
P/2008 J2 Beshore	2008 04 24.018	2.45485	0.27988
C/2008 J3 McNaught	2008 04 03.968	3.53197	1.00000
C/2008 J4 McNaught	2008 06 19.441	0.44485	1.00000
C/2008 J5 (Garradd)	2008 04 6.257	1.99115	1.00000
C/2008 J6 (Hill)	2008 04 11.338	2.00562	1.00000
C/2008 L2 (Hill)	2008 07 26.315	2.51073	1.00000
C/2008 L3 (Hill)	2008 04 22.262	2.01103	1.00000
C/2008 N1 (Holmes)	2009 09 25.227	2.78329	1.00000
P/2008 O2 (McNaught)	2009 04 20.358	3.79833	0.15711
P/2008 O3 (Boattini)	2008 06 3.882	2.50275	0.70326
C/2008 P1 (Garradd)	2009 07 19.989	3.84722	1.00000
C/2008 Q1 (Maticic)	2008 12 29.960	2.96020	1.00000

```
Name (nome)                  Perihelium         q        e
P/2008 Q2 (Ory)              2008 10 19.0081 1.382229 0.573436
C/2008 Q3 (Garradd)          2009 06 23.0859 1.799265 1.000000
P/2008 QP20 (LINEAR-Hill)    2008 11  2.5188 1.723315 0.506178
P/2008 R1 (Garradd)          2008 07 25.2959 1.793056 0.342367
C/2008 R3 (LINEAR)           2008 11 22.477  1.90899  0.89605
P/2008 S1 (Catalina-McNaught) 2008 10  1.8160 1.190519 0.667118
C/2008 S3 (Boattini)         2011 06  5.693  8.01270  1.00000
P/2008 T1 (Boattini)         2008 02 26.8525 3.044840 0.280422
C/2008 T2 (Cardinal)         2009 06 12.847  1.20026  1.00000
P/2008 T4 (Hill)             2008 12 21.429  2.51086  0.43382
C/2008 X3 (LINEAR)           2008 10 11.026  1.91368  1.00000
P/2008 Y1 (Boattini)         2009 02 25.1897 1.270613 0.732532
P/2008 Y2 (Gibbs)            2009 01 22.4208 1.638386 0.543526
P/2008 Y3 (McNaught)         2009 01 11.9679 4.434222 0.447555
P/2009 B1 (Boattini)         2009 02  6.2408 2.426329 0.637146
C/2009 B2 (LINEAR)           2009 03  7.3216 2.327276 0.942085
C/2009 E1 (Itagaki)          2009 04  7.384  0.60809  1.00000
C/2009 F1 (Larson)           2009 06 24.835  1.83643  1.00000
C/2009 F2 (McNaught)         2009 11 27.403  5.85897  1.00000
C/2009 F4 (McNaught)         2011 12 31.003  5.41791  1.00000
C/2009 F5 (McNaught)         2008 11  9.983  2.31560  1.00000
C/2009 F6 (Yi-SWAN)          2009 05  7.451  1.27427  1.00000
C/2009 G1 (STEREO)           2009 04 16.806  1.12818  1.00000
P/2009 K1 (Gibbs)            2009 06 25.933  1.32261  0.63948
C/2009 K2 (Catalina)         2010 02  7.607  3.24439  1.00000
C/2009 K3 (Beshore)          2011 01  9.266  3.90156  1.00000
C/2009 K4 (Gibbs)            2009 06 18.950  1.55854  1.00000
C/2009 K5 (McNaught)         2010 04 30.073  1.42146  1.00000
P/2009 L2 (Yang-Gao)         2009 05 21.711  1.29714  0.62456
C/2009 O2 Catalina           2010 03 25.360  0.70903  1.00000
C/2009 O3 Hill               2009 05 10.972  2.47748  1.00000
C/2009 O4 Hill               2010 01 01.005  2.56811  1.00000
C/2009 P1 Garradd            2011 12 25.1787 1.556280 1.000000
C/2009 P2 Boattini           2010 02 10.8784 6.543817 1.001834
P/2009 Q1 Hill               2009 07 04.1083 2.788695 0.496536
P/2009 Q4 Boattini           2009 11 20.185  1.31873  0.57427
P/2009 Q5 McNaught           2009 09 08.490  2.91897  0.60955
C/2009 R1 McNaught           2010 07 02.0338 0.401355 1.000000
P/2009 S1 Gibbs              2009 08 02.782  2.41458  0.36140
P/2009 S2 McNaught           2009 06 23.6201 2.203599 0.470488
C/2009 S3 Lemmon             2011 12 25.349  6.68049  1.00000
P/2009 SK280 Spacewatch-Hill 2009 05 30.5343 4.207578 0.118092
C/2009 T1 McNaught           2009 10 02.154  6.23039  1.00000
P/2009 T2 La Sagra           2010 01 12.8216 1.754716 0.768630
C/2009 T3 LINEAR             2010 01 11.963  2.28179  1.00000
C/2009 U1 Garradd            2010 07 07.819  2.96357  1.00000
C/2009 U3 Hill               2010 03 18.531  1.37750  1.00000
P/2009 U4 McNaught           2009 09 11.164  1.67495  0.74565
C/2009 U5 Grauer             2010 08 23.034  0.57126  1.00000
P/2009 U6 LINEAR             2009 08 08.6607 1.484203 0.561147
P/2009 W1 Hill               2009 09 10.190  2.93817  0.32723
C/2009 W2 Boattini           2010 03 23.937  6.95031  1.00000
C/2009 Y1 Catalina           2011 01 25.403  2.44938  1.00000
C/2009 Y2 Kowalski           2010 04 02.849  2.29149  1.00000
P/2010 A1 Hill               2009 08 07.046  1.95063  0.55259
P/2010 A2 LINEAR             2009 12 03.267  2.00593  0.12433
P/2010 A3 Hill               2010 04 03.6800 1.621786 0.732187
C/2010 A4 Siding Spring      2010 10 07.770  2.74078  1.00000
P/2010 A5 LINEAR             2010 04 21.027  1.69746  0.63197
C/2010 B1 Cardinal           2011 02 07.1332 2.942013 0.998876
P/2010 B2 WISE               2009 12 20.8910 1.606972 0.463138
P/2010 C1 Scotti             2009 12 01.2703 5.235048 0.258847
```

15

```
Name (nome)               Perihelium          q         e
P/2010 D1 WISE            2009 06 25.6869  2.669078  0.356589
P/2010 D2 WISE            2010 02 27.010   3.65735   0.44667
C/2010 D3 WISE            2010 09 03.8726  4.246771  0.999681
C/2010 D4 WISE            2009 03 30.2313  7.146656  0.888703
C/2010 DG56 WISE          2010 05 13.5729  1.612228  1.000000
C/2010 E1 Garradd         2009 11 09.622   2.67863   1.00000
P/2010 E2 Jarnac          2010 04 07.9999  2.398966  0.722511
C/2010 E3 WISE            2010 04 04.326   2.27420   1.00000
C/2010 E5 Scotti          2011 03 15.739   4.34469   1.00000
C/2010 F1 Boattini        2009 11 10.4756  3.586869  0.946971
C/2010 F3 Scotti          2010 08 03.7226  5.444714  0.912726
C/2010 F4 (Machholz)      2010 04  6.109   0.61383   1.00000
C/2010 FB87 WISE-Garradd  2010 11 07.3239  2.843425  0.990077
C/2010 G1 (Boattini)      2010 04  2.527   1.20501   1.00000
C/2010 G2 Hill            2011 09 02.0415  1.980547  0.979545
C/2010 G3 WISE            2010 04 11.1237  4.907518  0.998631
C/2010 H1 (Garradd)       2010 07  4.167   2.63536   1.00000
P/2010 H2 Vales           2010 03 07.8208  3.105658  0.192470
P/2010 H4 (Scotti)        2012 01 17.085   1.42143   1.00000
P/2010 H5 Scotti          2010 04 11.1623  6.023288  0.155515
C/2010 J1 Boattini        2010 02 04.832   1.69575   0.95398
C/2010 J2 McNaught        2010 06 04.0284  3.387423  0.999816
P/2010 J3 McMillan        2010 08 23.5692  2.454691  0.727177
C/2010 J4 WISE            2010 05 03.171   1.08553   1.00000
P/2010 J5 McNaught        2009 11 02.8683  3.747623  0.086815
P/2010 JC81 WISE          2011 04 26.5607  1.811074  0.777368
P/2010 K2 WISE            2010 07 07.624   1.19683   0.58470
C/2010 KW7 WISE           2010 10 11.2641  2.571659  0.973685
P/2010 L1 WISE            2010 02 06.242   2.14890   0.47642
C/2010 L3 Catalina        2010 11 10.1353  9.882410  1.001218
C/2010 L4 WISE            2010 03 23.803   2.95042   1.00000
C/2010 L5 WISE            2010 04 23.8414  0.800821  0.921681
C/2010 M1 Gibbs           2012 02 07.840   2.29869   1.00000
P/2010 N1 WISE            2010 08 16.1493  1.494521  0.533842
P/2010 P4 WISE            2010 07 06.4827  1.862466  0.514710
C/2010 R1 LINEAR          2012 05 18.9628  5.621394  1.003687
P/2010 R2 La Sagra        2010 06 23.5762  2.619202  0.153719
C/2010 S1 LINEAR          2013 05 20.3053  5.900221  1.000625
P/2010 T1 McNaught        2010 10 29.8249  3.212654  0.314502
P/2010 T2 PANSTARRS       2011 07 10.7831  3.752845  0.318421
P/2010 TO20 LINEAR-Grauer 2008 11 05.2255  5.201784  0.079340
P/2010 U1 Boattini        2010 03 24.7956  4.903076  0.257290
P/2010 U2 Hill            2010 11 09.4117  2.553738  0.402477
C/2010 U3 Boattini        2019 02 26.3523  8.467982  1.001123
P/2010 UH55 Spacewatch    2011 05 10.5041  2.768311  0.575471
P/2010 V1 Ikeya-Murakami  2010 10 13.0944  1.576500  0.486346
C/2010 V1 Ikeya-Murakami  2010 10 26.513   1.74377   1.00000
P/2010 WK LINEAR          2010 10 19.8282  1.766245  0.692144
C/2010 X1 Elenin          2011 09 10.7351  0.482206  0.999970
P/2011 A2 Scotti          2010 12 24.4439  1.586390  0.563625
C/2011 A3 Gibbs           2011 12 16.0818  2.344863  0.997701
C/2011 C1 McNaught        2011 04 17.9588  0.883752  1.000000
P/2011 C2 Gibbs           2012 01 07.6340  5.388580  0.269017
C/2011 C3 Gibbs           2011 04 07.4846  1.517927  1.000000
C/2011 F1 LINEAR          2013 01 08.0274  1.818765  1.000038
C/2011 G1 McNaught        2011 09 16.4208  2.155254  1.000936
C/2011 J2 LINEAR          2013 12 25.2466  3.445219  1.000402
C/2011 J3 LINEAR          2011 01 24.2652  1.449067  0.924171
P/2011 JB15 Spacewatch-Boattin2012 01 21.0194  5.019191  0.318023
C/2011 K1 Schwartz-Holvorcem  2011 04 20.9245  3.368397  0.981487
C/2011 L1 McNaught        2010 12 18.2853  2.241790  0.796789
C/2011 L2 McNaught        2011 11 01.2726  1.943488  1.001792
```

16

```
Name (nome)                     Perihelium         q          e
C/2011 L3 McNaught              2011 08 10.5020 1.923901 0.999987
C/2011 L4 PANSTARRS             2013 03 10.1698 0.301542 1.000031
C/2011 M1 LINEAR                2011 09 07.5791 0.895651 1.002430
P/2011 N1 ASH                   2012 05 31.1988 2.857934 0.546227
C/2011 N2 McNaught              2011 10 18.7332 2.563299 0.999281
P/2011 NO1                      2011 01 21.4166 1.244934 0.774737
C/2011 O1 LINEAR                2012 08 18.4640 3.890543 0.997048
P/2011 P1 McNaught              2010 07 17.7352 4.975204 0.342221
C/2011 P2 PANSTARRS             2010 09 12.6342 6.146991 0.369982
C/2011 Q1 PANSTARRS             2011 06 30.6453 6.780403 0.998036
C/2011 Q2 McNaught              2012 01 19.7947 1.349615 1.000000
P/2011 Q3 McNaught              2011 08 14.3226 2.367437 0.532347
C/2011 Q4 SWAN                  2011 09 21.0774 1.112110 0.974050
C/2011 R1 McNaught              2012 10 19.6169 2.079681 1.000572
P/2011 R2 PANSTARRS             2011 11 24.1873 2.049235 0.412690
P/2011 R3 Novichonok-Gerke      2012 04 02.9653 3.557918 0.267049
P/2011 S1 Gibbs                 2014 08 07.6053 7.192431 0.136767
C/2011 S1 Gibbs                 2013 01 16.5374 4.987846 1.000000
C/2011 S2 Kowalski              2011 10 26.3522 1.115130 0.931657
P/2011 U1 PANSTARRS             2012 06 20.6359 2.356191 0.417754
P/2011 U2 Bressi                2012 05 06.9441 4.837431 0.112941
C/2011 U3 PANSTARRS             2012 06 03.9296 1.071370 1.000052
P/2011 UA134 Spacewatch-PANSTA2011 12 07.1964 2.051428 0.632423
C/2011 UF305 LINEAR             2012 07 22.1547 2.138225 1.000637
P/2011 V1 Boattini              2011 05 12.1831 1.711668 0.551535
P/2011 VJ5 Lemmon               2011 12 08.6891 1.505780 0.555396
P/2011 W1 PANSTARRS             2012 01 23.5641 3.310933 0.288734
P/2011 W2 Rinner                2011 11 06.2220 2.303108 0.393708
C/2011 W3 Lovejoy               2011 12 16.0113 0.005557 0.999928
P/2011 Y1 Levy                  2012 01 14.9364 1.007454 0.668259
P/2011 Y2 Boattini              2012 03 21.6913 1.787188 0.712729
C/2011 Y3 Boattini              2011 08 26.5540 3.495488 0.701036
C/2012 A1 PANSTARRS             2013 11 29.0428 7.606360 1.000000
C/2012 A2 LINEAR                2012 11 04.4379 3.540152 1.000000
P/2012 B1 PANSTARRS             2013 07 20.3347 3.809343 0.415867
C/2012 B3 La Sagra              2011 11 29.9102 3.527273 1.000000
C/2012 C1 McNaught              2013 03 01.1677 4.629673 1.000000
C/2012 C2 Bruenjes              2012 03 12.8472 0.801913 1.000000
P/2012 C3 PANSTARRS             2011 10 06.7469 3.605925 0.630111
C/2012 CH17 MOSS                2012 09 28.1824 1.295711 1.000000
C/2012 E1 Hill                  2011 05 04.7364 7.402674 1.000000
C/2012 E3 PANSTARRS             2011 05 27.2685 3.953505 1.000000
C/2012 F1 Gibbs                 2012 02 19.2211 2.556243 1.000000
P/2012 F2 PANSTARRS             2013 04 07.9357 2.731989 0.623660
C/2012 F3 PANSTARRS             2015 04 05.3989 3.468334 1.000000
P/2012 F5 Gibbs                 2010 03 20.0774 2.872637 0.043305
C/2012 F6 Lemmon                2013 03 24.5138 0.731247 0.998534
P/2012 G1 PANSTARRS             2012 06 01.4325 2.584166 0.380722
P/2012 H1 PANSTARRS             2011 03 19.5286 3.479974 0.209434
C/2012 H2 McNaught              2012 05 03.7006 1.714926 0.883358
C/2012 J1 Catalina              2012 11 28.9475 3.220878 1.000000
C/2012 K1 PANSTARRS             2014 09 27.8301 1.246557 1.000000
P/2012 K3 Gibbs                 2012 10 08.5460 2.031492 0.452610
C/2012 K5 LINEAR                2012 11 29.0439 1.154235 1.000000
C/2012 K6 McNaught              2013 05 21.1804 3.366320 1.000000
C/2012 K8 Lemmon                2014 08 23.1468 6.426101 1.000000
C/2012 L1 LINEAR                2012 12 25.7035 2.257710 0.994746
C/2012 L2 LINEAR                2013 05 08.1987 1.498592 1.000000
C/2012 L3 LINEAR                2012 06 11.5790 3.048323 1.000000
P/2012 NJ La Sagra              2012 06 13.0902 1.291891 0.848056
P/2012 O1 McNaught              2012 07 23.9131 1.498896 0.578513
P/2012 O2 McNaught              2012 06 25.1015 1.660826 0.538209
```

Name (nome)	Perihelium	q	e
P/2012 O3 McNaught	2012 08 16.0613	1.599314	0.648597
C/2012 Q1 Kowalski	2012 06 15.8307	9.504756	0.637093
C/2012 S1 ISON	2013 11 28.8337	0.012518	1.000002
P/2012 S2 La Sagra	2012 08 18.9660	1.374669	0.700906
C/2012 S3 PANSTARRS	2013 09 03.6446	2.598798	1.000000
C/2012 S4 PANSTARRS	2013 07 12.0434	4.294175	0.886382
P/2012 SB6 Lemmon	2012 10 31.3723	2.406080	0.384563
P/2012 T1 PANSTARRS	2012 11 21.9160	2.419162	0.206284
P/2012 T2 PANSTARRS	2011 08 29.4442	4.884726	0.114565
P/2012 T3 PANSTARRS	2012 04 30.3931	2.407888	0.596998
C/2012 T4 McNaught	2012 10 10.3863	1.960445	1.000000
C/2012 T5 Bressi	2013 02 24.0595	0.322813	1.000403
C/2012 T6 Kowalski	2012 08 25.0896	1.794971	1.000000
P/2012 T7 Vorobjov	2012 06 16.5841	3.785996	0.336050
P/2012 TK8 Tenagra	2013 05 09.7986	3.089703	0.261374
P/2012 TB36 Lemmon	2013 07 18.7558	4.266327	0.384393
C/2012 U1 PANSTARRS	2014 07 17.1153	5.690924	1.000000
P/2012 U2 PANSTARRS	2012 08 27.8394	3.452330	0.558845
P/2012 US27 Siding Spring	2013 02 08.5770	1.820649	0.648233
C/2012 V1 PANSTARRS	2013 07 21.7076	2.085595	0.998065
C/2012 V2 LINEAR	2013 08 16.5477	1.454598	0.997180
C/2012 X1 LINEAR	2014 02 16.1414	1.580173	1.007296
C/2012 X2 PANSTARRS	2013 04 03.3296	4.735371	0.768570
C/2012 Y1 LINEAR	2012 12 12.9442	2.128507	1.000000
C/2012 Y3 McNaught	2012 08 25.8209	1.764345	0.939654
C/2013 A1 Siding Spring	2014 10 25.0469	1.395790	1.000362
P/2013 A2 Scotti	2013 02 08.3109	2.179255	0.456117
P/2013 AL76 Catalina	2012 12 13.4962	2.046881	0.674470
C/2013 B2 Catalina	2013 06 30.5906	3.753316	1.000000
C/2013 C2 Tenagra	2010 08 31.0251	8.962433	0.498788
P/2013 CE31 MOSS	2012 05 17.8500	4.014682	0.172874
C/2013 D1 Holvorcem	2013 04 08.8521	2.485151	0.793213
C/2013 E1 McNaught	2013 03 07.6620	7.809776	1.000000
C/2013 E2 Iwamoto	2013 03 08.0465	1.392327	0.920413
C/2013 F1 Boattini	2012 12 08.0155	1.933769	1.000000
C/2013 F2 Catalina	2013 04 18.7722	6.217827	0.999210
C/2013 F3 McNaught	2013 05 25.3073	2.251821	0.994701
C/2013 G1 Kowalski	2013 11 24.1689	3.404195	0.495268
C/2013 G2 McNaught	2012 12 05.8180	2.144402	1.008854
C/2013 G3 PANSTARRS	2014 11 10.4509	3.796460	1.000000
P/2013 G4 PANSTARRS	2013 01 31.4542	2.603657	0.420144
C/2013 G5 Catalina	2013 09 01.6322	0.929163	1.000139
C/2013 G6 Lemmon	2013 07 24.8244	2.050468	1.004067
C/2013 G7 McNaught	2014 03 16.8441	4.690698	1.000000
C/2013 G8 PANSTARRS	2013 11 27.3382	5.111244	1.005077
C/2013 G9 Tenagra	2015 01 30.5163	5.036790	1.000000
C/2013 H1 La Sagra	2013 06 01.3726	2.622218	1.000000
C/2013 H2 Boattini	2011 03 27.7999	2.481260	0.774571
P/2013 J2 McNaught	2013 08 26.9872	2.086849	0.503148
C/2013 J3 McNaught	2013 05 23.8455	4.038172	1.000000
P/2013 J4 PANSTARRS	2013 07 24.8517	2.287672	0.646077

Name (nome)	w	Ω	I	G	H
1P/Halley	111.3000	58.4000	162.3000	5.0	10.0
2P/Encke	186.5358	334.5730	11.7789	11.5	6.0
3D/Biela	221.7000	250.7000	13.2000	5.0	10.0
4P/Faye	205.0119	199.2925	9.0495	8.0	6.0
5D/Brorsen	14.9000	103.0000	29.4000	5.0	10.0
6P/d'Arrest	178.1395	138.9423	19.5141	7.5	16.0
7P/Pons-Winnecke	172.3099	93.4241	22.3105	10.0	6.0
8P/Tuttle	207.5190	270.3435	54.9831	8.0	8.0
9P/Tempel	179.0726	68.8819	10.5247	5.5	10.0
10P/Tempel	195.5471	117.8024	12.0269	5.0	10.0
11P/Tempel-Swift-LINEAR	163.8984	240.5262	13.5532	17.0	4.0
12P/Pons-Brooks	199.0000	255.9000	74.2000	5.0	10.0
13P/Olbers	64.6000	86.1000	44.6000	5.0	10.0
14P/Wolf	158.7057	202.3569	27.9961	5.5	12.0
15P/Finlay	347.4932	13.8819	6.8230	12.0	4.0
16P/Brooks	219.6352	159.3219	4.2582	7.5	10.0
17P/Holmes	24.4805	326.8038	19.0919	10.0	6.0
18D/Perrine-Mrkos	166.1000	240.9000	17.8000	5.0	10.0
19P/Borrelly	353.3298	75.4478	30.3255	4.5	10.0
20D/Westphal	57.1000	348.0000	40.9000	5.0	10.0
21P/Giacobini-Zinner	172.6022	195.3970	31.9105	9.0	6.0
22P/Kopff	162.7903	120.8979	4.7239	3.0	10.4
23P/Brorsen-Metcalf	129.6000	311.6000	19.3000	5.0	10.0
24P/Schaumasse	58.0235	79.7161	11.7297	6.5	14.0
25D/Neujmin 2	10.6387	193.7040	328.7169	5.0	10.0
26P/Grigg-Skjellerup	2.1029	211.5680	22.4262	12.0	16.0
27P/Crommelin	195.9920	250.6336	28.9585	12.0	8.0
28P/Neujmin	347.5837	346.3993	14.3503	8.5	6.0
29P/Schwassmann-Wachmann	50.4735	312.5717	9.3791	4.0	4.0
30P/Reinmuth	13.1574	119.7312	8.1246	9.5	6.0
31P/Schwassmann-Wachmann	17.9709	114.1690	4.5460	5.0	8.0
32P/Comas Sola	53.2804	57.8634	9.9711	6.5	8.0
33P/Daniel	18.9580	66.5621	22.3747	10.0	12.0
34D/Gale	11.7281	209.1572	67.9235	5.0	10.0
35P/Herschel-Rigollet	64.2070	29.29800	355.9800	5.0	10.0
36P/Whipple	201.5426	182.3752	9.9262	8.5	6.0
37P/Forbes	329.3894	315.0309	8.9559	10.5	4.8
38P/Stephan-Oterma	17.9814	358.1910	79.1883	5.0	10.0
39P/Oterma	1.9428	56.2783	331.5811	5.0	10.0
40P/Vaisala 1	11.5384	47.1894	134.7325	5.0	10.0
41P/Tuttle-Giacobini-Kresak	62.1704	141.0618	9.2252	10.0	16.0
42P/Neujmin 3	3.9854	147.1604	150.3839	5.0	10.0
43P/Wolf-Harrington	191.5094	249.8741	15.9727	8.0	6.0
44P/Reinmuth	58.3436	286.4771	5.8951	8.3	6.0
45P/Honda-Mrkos-Pajdusakova	326.2530	89.0030	4.2523	13.5	8.0
46P/Wirtanen	356.3182	82.1602	11.7563	9.0	6.0
47P/Ashbrook-Jackson	357.9173	357.0026	13.0419	1.0	11.2
48P/Johnson	207.9938	117.2635	13.6612	10.0	6.0
49P/Arend-Rigaux	332.7978	118.8746	19.0497	11.3	4.4
50P/Arend	49.0457	355.3373	19.1559	9.5	6.0
51P/Harrington	269.1930	83.7652	5.4254	11.5	8.0
52P/Harrington-Abell	139.6326	336.8638	10.2301	13.5	6.0
53P/Van Biesbroeck	6.6099	134.0832	149.0022	5.0	10.0
54P/de Vico-Swift-NEAT	1.9406	358.8616	6.0670	10.0	6.0
55P/Tempel-Tuttle	162.4865	172.5002	235.2709	5.0	10.0
56P/Slaughter-Burnham	8.1556	44.0779	346.2636	5.0	10.0
57P/du Toit-Neujmin-Delporte	115.3152	188.8430	2.8482	12.5	6.0
58P/Jackson-Neujmin	200.4886	160.6350	13.4899	15.5	6.0
59P/Kearns-Kwee	127.5540	313.0385	9.3413	7.0	6.0
60P/Tsuchinshan	216.3774	267.6840	3.6105	11.5	6.0
61P/Shajn-Schaldach	221.8977	163.0560	6.0055	6.0	10.0
62P/Tsuchinshan	30.2431	90.3032	9.7132	8.0	10.0

Name (nome)	W	Ω	I	G	H
63P/Wild	168.9634	358.0240	19.7807	10.5	6.0
64P/Swift-Gehrels	96.3046	300.7414	8.9514	8.5	12.0
65P/Gunn	198.8081	67.8776	10.4793	5.0	6.0
66P/du Toit	18.7007	257.2497	22.2153	5.0	10.0
67P/Churyumov-Gerasimenko	11.5483	50.7902	7.1064	11.0	4.0
68P/Klemola	154.0362	175.3631	11.1461	10.0	4.0
69P/Taylor	343.4548	104.8780	22.0460	9.5	12.0
70P/Kojima	2.1148	119.2608	6.5953	11.0	6.0
71P/Clark	208.8225	59.6079	9.4812	9.8	6.0
72D/Denning-Fujikawa	8.6431	334.3171	41.5334	5.0	10.0
73P/Schwassmann-Wachmann	198.8732	69.8424	11.3793	11.5	6.0
74P/Smirnova-Chernykh	86.6846	77.0819	6.6508	5.0	6.0
75D/Kohoutek	5.9072	175.8017	269.6860	5.0	10.0
76P/West-Kohoutek-Ikemura	0.0702	84.1243	30.4830	8.0	12.0
77P/Longmore	197.0427	14.9139	24.3572	7.0	8.0
78P/Gehrels	192.7768	210.5585	6.2552	5.5	8.0
79P/du Toit-Hartley	281.6693	280.6421	3.1453	16.0	4.0
80P/Peters-Hartley	338.5934	259.8818	29.8974	9.0	8.0
81P/Wild	41.7329	136.0993	3.2376	7.0	6.0
82P/Gehrels	226.5374	239.4115	1.1267	5.0	8.0
83D/Russell 1	22.6594	0.3762	230.8420	5.0	10.0
84P/Giclas	276.4281	112.3850	7.2851	9.5	8.0
85P/Boethin	53.5433	343.4713	4.2157	6.5	4.0
86P/Wild	179.1612	72.5823	15.4487	11.0	6.0
87P/Bus	24.5824	181.9235	2.6001	7.2	10.0
88P/Howell	235.8418	56.7440	4.3826	11.0	6.0
89P/Russell	249.2856	42.3958	12.0315	11.5	6.0
90P/Gehrels 1	9.6152	28.2375	13.5166	5.0	10.0
91P/Russell	354.7068	247.8741	14.0753	7.5	6.0
92P/Sanguin	18.7646	163.0704	182.3399	5.0	10.0
93P/Lovas	74.6697	339.9252	12.2186	9.5	6.0
94P/Russell	92.8775	70.9216	6.1829	9.0	6.0
95P/Chiron	339.0073	209.3696	8.4334	6.5	2.0
96P/Machholz	14.7561	94.3243	58.2988	13.0	4.8
97P/Metcalf-Brewington	228.2016	185.2008	17.8855	5.5	6.0
98P/Takamizawa	157.9032	114.7439	10.5439	9.0	8.0
99P/Kowal	172.8654	28.4003	4.3443	4.5	6.0
100P/Hartley	181.7049	37.8476	25.6527	9.0	8.0
101P/Chernykh	5.0793	263.1723	130.2710	5.0	10.0
102P/Shoemaker	18.7379	339.8769	26.2453	6.5	8.0
103P/Hartley 2	181.2028	219.7602	13.6184	8.7	20.0
104P/Kowal	200.4330	235.6319	10.2706	12.5	4.0
105P/Singer Brewster	46.6702	192.4185	9.1706	11.5	6.0
106P/Schuster	355.8289	50.6122	20.1113	14.0	4.8
107P/Wilson-Harrington	91.3117	270.5228	2.7843	15.5	2.0
108P/Ciffreo	357.9801	53.7389	13.0784	8.0	12.0
109P/Swift-Tuttle	113.4538	152.9821	139.3811	5.0	10.0
110P/Hartley	167.6336	287.7082	11.6922	1.0	12.0
111P/Helin-Roman-Crockett	4.4768	89.9430	4.2090	5.0	8.0
112P/Urata-Niijima	21.4495	31.9274	24.2036	14.0	6.0
113P/Spitaler	49.8685	14.4736	5.7760	13.5	4.0
114P/Wiseman-Skiff	172.8584	271.0564	18.2836	11.5	6.0
115P/Maury	120.0736	176.6026	11.7059	10.5	6.0
116P/Wild	173.6000	21.0512	3.6108	2.5	10.0
117P/Helin-Roman-Alu	222.9078	58.8974	8.6991	2.5	8.0
118P/Shoemaker-Levy	301.8459	151.8702	8.4880	12.0	4.0
119P/Parker-Hartley	181.1703	244.0995	5.1960	3.5	8.0
120P/Mueller	30.0850	4.4527	8.7977	12.0	4.0
121P/Shoemaker-Holt	13.7149	94.2757	20.1089	6.5	8.0
122P/de Vico	85.3827	12.9960	79.6245	5.0	10.0
123P/West-Hartley	102.8221	46.5984	15.3572	4.0	10.0
124P/Mrkos	183.6830	0.5204	31.5566	13.5	2.8

Name (nome)	w	Ω	I	G	H
125P/Spacewatch	87.2403	153.1906	9.9863	13.0	6.0
126P/IRAS	356.7469	357.7654	45.8278	6.0	8.0
127P/Holt-Olmstead	6.5239	13.6877	14.3194	11.0	6.0
128P/Shoemaker-Holt	210.3606	214.5790	4.3626	8.5	4.0
129P/Shoemaker-Levy	305.4152	187.2454	3.5028	11.0	4.0
130P/McNaught-Hughes	224.4565	89.7896	7.3083	10.0	6.0
131P/Mueller	179.4806	214.2186	7.3559	11.0	4.0
132P/Helin-Roman-Alu 2	5.7659	221.1010	178.3881	5.0	10.0
133P/Elst-Pizarro	132.1895	160.1535	1.3868	15.4	2.0
134P/Kowal-Vavrova	18.5103	202.1358	4.3485	11.5	4.0
135P/Shoemaker-Levy	21.7252	213.1338	6.0617	7.0	8.0
136P/Mueller	225.2537	137.5275	9.4193	11.0	4.0
137P/Shoemaker-Levy	141.3004	233.3510	4.8949	11.0	4.0
138P/Shoemaker-Levy	95.6248	309.4151	10.0837	15.0	6.0
139P/Vaisala-Oterma	166.1313	242.2735	2.3334	9.5	4.0
140P/Bowell-Skiff	3.8358	173.0867	343.4564	5.0	10.0
141P/Machholz	149.3454	246.0891	12.8038	12.0	12.0
142P/Ge-Wang	175.5943	176.5156	12.3067	8.5	6.0
143P/Kowal-Mrkos	320.8714	245.3285	4.6893	13.5	2.0
144P/Kushida	216.0994	245.5651	4.1093	8.5	8.0
145P/Shoemaker-Levy	10.0212	26.9485	11.2969	13.5	4.0
146P/Shoemaker-LINEAR	316.8314	53.5713	23.0796	15.0	4.0
147P/Kushida-Muramatsu	346.9245	93.7397	2.3673	14.0	4.0
148P/Anderson-LINEAR	6.7131	89.8070	3.6783	16.0	2.0
149P/Mueller	43.7897	145.2662	29.7354	8.0	8.0
150P/LONEOS	245.7067	272.4345	18.5002	13.5	4.0
151P/Helin	4.7156	215.3209	143.5608	5.0	10.0
152P/Helin-Lawrence	163.8109	91.9115	9.8670	11.5	4.0
153P/Ikeya-Zhang	28.1198	34.6731	93.3702	5.0	10.0
154P/Brewington	49.0070	343.5022	17.8304	2.5	12.0
155P/Shoemaker 3	6.3858	14.9109	97.2645	5.0	10.0
156P/Russell-LINEAR	357.7287	39.0000	20.7824	15.5	2.0
157P/Tritton	148.8459	300.0138	7.2846	10.0	4.0
158P/Kowal-LINEAR	232.3831	137.3083	7.9079	9.0	4.0
159P/LONEOS	4.7893	55.1044	23.4773	10.0	4.0
160P/LINEAR	18.1823	337.0028	17.2755	15.5	2.0
161P/Hartley-IRAS	95.7081	47.1765	1.4649	5.0	10.0
162P/Siding Spring	356.4013	31.2365	27.8106	12.0	4.0
163P/NEAT	349.6364	102.1180	12.7168	14.5	4.0
164P/Christensen	325.8688	88.3247	16.2608	11.0	4.0
165P/LINEAR	15.9052	126.2053	0.6387	5.0	10.0
166P/NEAT	322.0842	64.3619	15.3604	5.5	4.0
167P/CINEOS	344.6641	295.8565	19.1000	9.5	2.0
168P/Hergenrother	13.9542	356.4707	21.9294	15.5	4.0
169P/NEAT	218.0102	176.1639	11.2968	16.0	2.0
170P/Christensen	10.1277	225.3842	143.0124	5.0	10.0
171P/Spahr	347.0827	101.7210	21.9481	13.5	4.0
172P/Yeung	178.9153	40.1040	11.5178	13.0	4.0
173P/Mueller	29.3933	100.4957	16.5059	7.5	4.0
174P/Echeclus	162.9001	173.3916	4.3415	9.4	2.0
175P/Hergenrother	55.9728	123.5906	6.0779	14.0	4.0
176P/LINEAR	35.7092	346.0712	0.2348	15.1	2.0
177P/Barnard	31.2178	60.4572	272.0666	5.0	10.0
178P/Hug-Bell	296.9324	103.5743	10.9750	13.5	4.0
179P/Jedicke	295.4862	115.8579	19.8743	2.5	8.0
180P/NEAT	94.8901	84.7555	16.9135	11.0	4.0
181P/Shoemaker-Levy	333.7775	37.6931	16.9913	11.5	4.0
182P/LONEOS	53.7617	72.8683	16.2493	18.0	4.0
183P/Korlevic-Juric	160.7839	5.8352	18.7432	12.5	4.0
184P/Lovas	78.0719	277.7313	1.5515	13.5	4.0
185P/Petriew	181.9418	214.0912	14.0077	15.0	4.0
186P/Garradd	287.4906	327.4976	28.5142	7.5	4.0

21

Name (nome)	w	Ω	I	G	H
187P/LINEAR	136.8504	111.1942	13.6302	9.0	4.0
188P/LINEAR-Mueller	26.4343	359.1496	10.5468	11.5	4.0
189P/NEAT	15.3478	282.1553	20.3745	19.0	4.0
190P/Mueller	49.7346	336.1176	2.1899	13.0	4.0
191P/McNaught	274.4524	106.4085	8.7629	13.0	4.0
192P/Shoemaker-Levy	312.8429	51.6504	24.5611	15.0	4.0
193P/LINEAR-NEAT	8.2744	335.2586	10.7033	14.0	4.0
194P/LINEAR	130.6398	352.0623	11.1186	16.0	4.0
195P/Hill	249.4930	243.1926	36.3902	8.5	4.0
196P/Tichy	11.7166	24.3410	19.3780	13.5	4.0
197P/LINEAR	188.7473	66.3895	25.5423	16.5	2.0
198P/ODAS	69.0749	358.4913	1.3412	12.5	4.0
199P/Shoemaker	191.9492	92.9466	24.7671	10.0	4.0
200P/Larsen	134.1208	234.7752	12.1146	9.0	4.0
201P/LONEOS	24.9633	35.2995	7.0324	14.0	4.0
202P/Scotti	255.5572	194.5826	2.1849	13.5	4.0
203P/Korlevic	154.7497	290.5130	2.9738	14.5	2.0
204P/LINEAR-NEAT	355.0388	109.1065	6.5812	14.0	4.0
205P/Giacobini	154.2271	179.6311	15.3038	13.0	4.0
206P/Barnard-Boattini	181.4287	204.1347	32.9285	19.0	4.0
207P/NEAT	271.1732	200.6739	10.1500	16.0	4.0
208P/McMillan	310.5032	36.4171	4.4134	12.5	4.0
209P/LINEAR	149.7234	66.4549	19.1480	17.0	2.0
210P/Christensen	345.7777	93.8869	10.2154	13.5	4.0
211P/Hill	4.4118	117.2959	18.8724	12.5	4.0
212P/NEAT	15.0488	98.9290	22.3979	17.0	2.0
213P/Van Ness	3.4408	312.6274	10.2362	10.5	4.0
214P/LINEAR	190.2702	348.2598	15.2135	13.0	4.0
215P/NEAT	224.3804	75.1146	12.8354	11.0	4.0
216P/LINEAR	151.5794	359.8942	9.0356	13.0	4.0
217P/LINEAR	246.7441	125.6223	12.8816	12.0	4.0
218P/LINEAR	10.6050	226.7437	18.1516	16.0	4.0
219P/LINEAR	107.6047	231.0283	11.5292	11.0	4.0
220P/McNaught	180.7742	150.1192	8.1325	15.0	4.0
221P/LINEAR	39.6991	230.0164	11.4178	14.5	4.0
222P/LINEAR	345.4306	7.1337	5.1474	20.0	4.0
223P/Skiff	37.8924	346.8286	27.0515	12.0	4.0
224P/LINEAR-NEAT	16.0359	40.5401	13.4334	15.5	4.0
225P/LINEAR	3.8242	14.2259	21.3958	18.0	4.0
226P/Pigott-LINEAR-Kowalski	340.9630	54.0678	44.0212	12.5	4.0
227P/Catalina-LINEAR	90.1393	49.8858	6.5248	16.5	2.0
228P/LINEAR	114.8059	31.0683	7.9151	14.5	2.0
229P/Gibbs	224.0181	157.9802	26.1104	13.0	4.0
230P/LINEAR	308.7323	112.5092	14.6426	13.0	4.0
231P/LINEAR-NEAT	42.4327	133.0844	12.3253	14.5	2.0
232P/Hill	53.4226	56.1415	14.6344	11.5	4.0
233P/La Sagra	27.1338	74.9974	11.2750	15.0	4.0
234P/LINEAR	358.3096	179.7281	11.5157	12.0	4.0
235P/LINEAR	333.7536	204.4847	8.8910	12.0	4.0
236P/LINEAR	119.3384	245.6746	16.3317	14.0	4.0
237P/LINEAR	20.9278	251.6609	16.4510	14.5	2.0
238P/Read	325.1383	51.6379	1.2655	14.5	4.0
239P/LINEAR	11.3166	220.3722	256.1164	5.0	10.0
240P/NEAT	352.0178	74.9615	23.5219	12.0	4.0
241P/LINEAR	110.1566	305.9731	20.8851	13.5	4.0
242P/Spahr	247.6717	180.7829	32.4850	8.0	4.0
243P/NEAT	283.7787	87.6721	7.6396	12.5	4.0
244P/Scotti	92.6123	354.1532	2.2591	9.0	4.0
245P/WISE	316.4414	318.5283	21.0858	14.0	4.0
246P/NEAT	176.2454	78.7840	15.9710	10.5	4.0
247P/LINEAR	47.3304	54.1254	13.6818	16.5	2.0
248P/Gibbs	209.9547	207.7838	6.3703	14.0	4.0

Name (nome)	w	Ω	I	G	H
249P/LINEAR	64.2276	240.4733	8.4251	15.5	4.0
250P/Larson	44.8947	73.7548	13.2930	14.5	4.0
251P/LINEAR	31.0047	219.4847	23.4933	16.5	2.0
252P/LINEAR	10.3810	343.2826	190.9962	5.0	10.0
253P/PANSTARRS	230.9830	146.9211	4.9386	14.5	4.0
254P/McNaught	220.6993	129.9503	32.5176	11.0	4.0
255P/Levy	179.6328	279.7444	18.2646	20.0	4.0
256P/LINEAR	124.1144	81.4464	27.6368	14.0	2.0
257P/Catalina	117.8128	207.8676	20.2448	11.5	4.0
258P/PANSTARRS	6.7462	25.8289	126.3448	5.0	10.0
259P/Garradd	256.5583	51.9623	15.8991	15.5	4.0
260P/McNaught	15.6896	351.9619	15.7357	13.5	4.0
261P/Larson	58.8050	298.4802	6.3254	14.0	4.0
262P/McNaught-Russell	171.1873	218.0120	29.0791	13.5	4.0
263P/Gibbs	14.4711	26.3155	113.3508	5.0	10.0
264P/Larsen	346.5183	220.9460	25.1198	13.0	4.0
265P/LINEAR	32.8570	344.7360	14.6895	14.5	4.0
266P/Christensen	98.0933	5.0967	3.4295	12.5	4.0
267P/LONEOS	97.2019	245.0124	5.3736	19.5	4.0
268P/Bernardi	15.6194	357.3621	129.7282	5.0	10.0
269P/Jedicke	223.5845	248.7660	6.6169	10.0	4.0
270P/Gehrels	211.0670	225.3037	2.8580	8.0	4.0
271P/van Houten-Lemmon	35.1205	9.5898	6.8584	11.0	4.0
272P/NEAT	27.8858	109.5025	18.1011	16.0	2.0
273P/Pons-Gambart	20.1696	320.4324	136.3977	11.5	4.0
274P/Tombaugh-Tenagra	38.4574	81.3626	15.8383	13.0	4.0
275P/Hermann	173.9845	348.7556	21.3427	15.0	4.0
276P/Vorobjov	205.7714	214.2925	14.4590	11.5	4.0
277P/LINEAR	152.3006	276.3629	16.7474	14.0	4.0
278P/McNaught	238.0031	15.5043	6.6822	14.0	4.0
279P/La Sagra	5.0546	5.9620	346.2607	5.0	10.0
280P/Larsen	104.6113	131.5136	11.7727	12.5	4.0
281P/MOSS	26.7639	87.1709	4.7238	11.0	4.0
C/1995 O1 (Hale-Bopp)	130.7820	282.8618	89.2480	-2.0	4.0
P/1998 QP54 (LONEOS-Tucker)	30.3333	341.8111	17.7115	15.0	2.0
P/1998 U4 (Spahr)	251.8496	181.6965	31.4835	8.0	4.0
P/1998 VS24 (LINEAR)	244.4171	159.1904	5.0271	13.0	2.0
P/1999 XN120 (Catalina)	161.7726	285.4562	5.0264	13.5	2.0
P/2000 Y3 (Scotti)	87.2495	354.7274	2.2518	9.0	4.0
C/2001 HT50 (LINEAR-NEAT)	323.9919	42.9384	163.2182	4.5	4.0
C/2001 K5 (LINEAR)	47.1691	237.4029	72.6095	4.0	4.0
C/2001 Q4 (NEAT)	1.2823	210.2894	99.6249	3.5	4.0
C/2002 J5 (LINEAR)	74.9642	314.0882	117.2552	11.0	2.0
C/2002 L9 (NEAT)	231.4541	110.4001	68.4451	8.5	2.0
C/2002 T7 (LINEAR)	158.0158	95.1039	160.5390	4.0	4.0
C/2002 VQ94 (LINEAR)	100.0689	35.0124	70.5180	9.5	2.0
C/2003 A2 (Gleason)	346.8396	154.5570	8.0673	3.5	4.0
C/2003 E1 (NEAT)	104.0013	137.0512	33.5094	12.5	2.0
C/2003 K4 (LINEAR)	198.4830	18.6908	134.2288	3.5	4.0
C/2003 T3 (Tabur)	43.7327	347.0542	50.5009	5.0	4.0
C/2003 T4 (LINEAR)	181.6680	93.8755	86.7898	6.0	4.0
C/2003 WT42 (LINEAR)	92.4878	48.4480	31.4119	9.0	2.0
P/2004 A1 (LONEOS)	21.8148	125.0017	10.5498	6.5	4.0
C/2004 B1 (LINEAR)	327.8905	272.8084	114.1026	10.5	2.0
C/2004 D1 (NEAT)	75.5616	62.2528	45.5380	11.5	2.0
P/2004 DO29 (Spacewatch-LINEAR	41.3028	147.5298	14.4842	13.5	2.0
P/2004 FY140 (LINEAR)	239.7555	327.1844	2.1271	12.5	2.0
C/2004 K1 (Catalina)	97.8025	326.9737	153.7396	7.0	4.0
C/2004 L2 (LINEAR)	257.3234	99.1676	62.5243	7.0	4.0
C/2004 Q2 (Machholz)	19.5567	93.4678	38.7192	5.5	4.0
C/2004 T3 (Siding Spring)	259.7257	50.3428	71.9707	2.5	4.0
P/2004 V5 (LINEAR-Hill)	88.0890	47.8536	19.3545	8.0	4.0

Name (nome)	w	Ω	I	G	H
P/2004 VR8 (LONEOS)	63.6333	71.1716	20.1531	10.0	4.0
C/2005 B1 (Christensen)	103.1872	195.5603	92.5445	6.5	4.0
C/2005 E2 (McNaught)	39.9601	347.8364	16.9918	5.5	4.0
C/2005 EL173 (LONEOS)	261.4938	344.7976	130.6798	11.5	2.0
C/2005 G1 (LINEAR)	113.8443	299.5799	108.4215	8.0	4.0
P/2005 JD108 (Catalina-NEAT)	90.3797	224.3003	3.2755	10.0	4.0
P/2005 JY126 (Catalina)	117.5978	207.9846	20.2344	11.5	4.0
C/2005 K1 (Skiff)	134.9617	106.3060	77.7566	7.0	4.0
P/2005 L1 (McNaught)	149.6248	138.3278	7.7358	9.5	4.0
C/2005 L3 (McNaught)	46.9587	288.6921	139.3950	4.0	4.0
C/2005 O1 (NEAT)	324.8940	304.8799	156.0114	13.5	2.0
C/2005 O2 (Christensen)	263.8396	280.7782	148.8911	9.5	4.0
C/2005 Q1 (LINEAR)	44.6668	87.7447	105.1953	6.0	4.0
C/2005 R4 (LINEAR)	6.8560	63.7737	164.0140	7.0	4.0
P/2005 RV25 (LONEOS-Christense	191.6688	246.9307	9.8837	9.5	4.0
P/2005 S2 (Skiff)	229.7592	161.2590	3.1405	7.5	4.0
C/2005 S4 (McNaught)	31.4463	318.2907	107.9597	5.0	4.0
P/2005 SB216 (LONEOS)	83.5722	1.6975	24.0981	12.0	2.0
C/2005 W2 (Christensen)	111.6726	336.6026	11.2664	9.5	4.0
P/2005 Y2 (McNaught)	194.6815	94.5974	19.1782	9.0	4.0
C/2005 YW (LINEAR)	234.6410	302.2114	40.5438	10.0	4.0
C/2006 A1 (Pojmanski)	351.2053	211.3523	92.7293	9.0	3.2
C/2006 CK10 (Catalina)	143.4468	243.8103	144.2629	14.5	2.0
C/2006 E1 (McNaught)	232.8067	95.0348	83.1944	6.0	4.0
P/2006 F1 (Kowalski)	186.4483	124.7898	21.2387	8.0	4.0
C/2006 F2 (Christensen)	181.2502	8.1428	20.4493	10.5	4.0
P/2006 G1 (McNaught)	313.9164	299.2448	18.5627	12.0	4.0
P/2006 HR30 (Siding Spring)	117.4132	309.9511	31.8848	11.0	4.0
C/2006 HW51 (Siding Spring)	359.9373	228.1246	45.8066	11.0	4.0
C/2006 K1 (McNaught)	296.4466	72.1152	53.8770	7.5	4.0
C/2006 K3 (McNaught)	328.0764	49.4031	92.6196	8.0	4.0
C/2006 K4 (NEAT)	233.6609	116.5963	111.3303	6.0	4.0
C/2006 L1 (Garradd)	338.4088	101.7615	143.2432	12.0	4.0
C/2006 L2 (McNaught)	48.0376	239.2460	101.0223	9.0	4.0
C/2006 M1 (LINEAR)	122.8920	231.6153	54.8770	8.5	4.0
C/2006 M4 (SWAN)	62.5935	148.7271	111.8226	8.5	4.0
C/2006 OF2 (Broughton)	95.6106	318.5094	30.1712	5.5	4.0
C/2006 P1 (McNaught)	155.9771	267.4146	77.8370	6.0	4.0
C/2006 Q1 (McNaught)	344.3555	199.5593	59.0439	5.0	4.0
P/2006 R2 (Christensen)	188.8848	139.1500	16.3155	11.0	4.0
C/2006 S2 (LINEAR)	166.3442	113.8865	98.9605	10.0	4.0
C/2006 S3 (LONEOS)	139.9563	38.3442	166.0319	2.0	4.0
P/2006 S4 (Christensen)	305.7259	36.1367	39.6299	11.0	4.0
C/2006 S5 (Hill)	182.1406	281.5572	10.1315	8.0	4.0
P/2006 T1 (Levy)	179.4635	279.8017	18.3217	10.5	4.0
P/2006 U5 (Christensen)	98.3885	5.0956	3.4304	12.5	4.0
C/2006 U6 (Spacewatch)	276.5989	180.1905	84.8831	8.0	4.0
C/2006 V1 (Catalina)	253.3997	335.7241	31.1190	8.0	4.0
C/2006 VZ13 (LINEAR)	174.1149	66.0267	134.7934	10.5	4.0
C/2006 W3 (Christensen)	133.4616	113.5594	127.0636	5.0	4.0
C/2006 WD4 (Lemmon)	292.6917	226.7906	152.7042	16.5	4.0
C/2006 X1 (LINEAR)	101.2914	255.2420	42.6181	7.0	4.0
C/2006 XA1 (LINEAR)	187.4194	318.6647	30.6292	10.0	4.0
P/2006 XG16 (Spacewatch)	41.3072	78.4540	9.0779	14.5	4.0
C/2006 YC (Catalina-Christense	335.5086	154.2874	69.5653	8.0	4.0
P/2007 B1 (Christensen)	46.9059	77.7859	12.3751	13.5	4.0
C/2007 B2 (Skiff)	205.9476	14.8850	27.4883	6.0	4.0
P/2007 C1 (Christensen)	95.6134	57.1156	8.0222	15.0	4.0
P/2007 C2 (Catalina)	179.4731	276.1188	8.6742	10.0	4.0
C/2007 D1 (LINEAR)	340.2429	171.1078	41.4358	3.5	4.0
C/2007 D3 (LINEAR)	309.0947	148.4337	45.9163	7.5	4.0
C/2007 E1 (Garradd)	7.1008	153.8268	174.3940	12.5	4.0

Name (nome)	w	Ω	I	G	H
C/2007 E2 (Lovejoy)	340.5460	232.4350	95.8884	9.5	4.0
C/2007 F1 (LONEOS)	153.7351	172.8731	116.0260	10.0	4.0
C/2007 G1 (LINEAR)	224.0622	78.9933	88.3591	5.5	4.0
P/2007 H1 (McNaught)	202.4726	144.3871	11.8716	10.0	4.0
P/2007 H3 (Garradd)	350.1519	263.6601	25.2146	14.0	4.0
C/2007 JA21 (LINEAR)	93.6135	65.5520	89.8266	8.0	4.0
C/2007 K1 (Lemmon)	51.569	294.757	108.492	6.0	4.0
P/2007 K2 (Gibbs)	345.725	189.810	7.623	14.0	4.0
C/2007 K3 (Siding Spring)	23.590	263.281	16.297	9.5	4.0
C/2007 K4 (Gibbs)	162.452	68.524	98.593	10.5	4.0
C/2007 K5 (Lovejoy)	255.576	193.737	64.885	12.0	4.0
C/2007 K6 (McNaught)	337.805	298.014	105.228	11.0	4.0
C/2007 M1 (McNaught)	53.404	326.822	139.750	6.0	4.0
C/2007 M2 (Catalina)	221.142	357.316	80.914	8.0	4.0
C/2007 M3 (LINEAR)	125.640	41.532	161.763	9.5	4.0
C/2007 N1 (McNaught)	266.796	115.220	9.329	12.0	4.0
C/2007 N3 Lulin	136.9164	338.4791	178.3704	6.5	4.0
C/2007 O1 LINEAR	159.3898	116.2338	24.3846	10.0	4.0
C/2007 P1 McNaught	89.602	186.747	118.855	13.0	4.0
C/2007 Q1 Garradd	281.837	5.932	81.871	10.5	4.0
P/2007 Q2 Gilmore	163.1530	172.2491	10.2384	16.0	4.0
C/2007 Q3 Siding Spring	2.1367	149.4394	65.6538	4.5	4.0
P/2007 R1 Larson	175.0699	181.6640	7.8910	8.0	4.0
P/2007 R2 Gibbs	353.0444	8.8249	1.4341	17.0	4.0
P/2007 R3 Gibbs	312.1308	30.0235	3.8145	13.5	4.0
P/2007 R4 Garradd	282.8669	87.5033	20.2260	14.5	4.0
P/2007 S1 Zhao	245.3891	141.6164	5.9725	13.0	4.0
C/2007 S2 Lemmon	211.2637	296.2435	16.8808	6.5	4.0
C/2007 T1 McNaught	233.725	111.429	117.648	10.5	4.0
P/2007 T2 Kowalski	358.560	3.999	9.895	18.5	4.0
P/2007 T4 Gibbs	42.006	37.038	23.864	13.0	4.0
C/2007 T5 Gibbs	34.507	109.847	45.623	8.0	4.0
P/2007 T6 Catalina	335.987	102.711	22.129	12.5	4.0
C/2007 U1 LINEAR	0.679	49.976	157.808	9.0	4.0
P/2007 V1 Larson	51.4466	8.1810	10.7890	12.0	4.0
P/2007 V2 Hill	278.5386	99.8672	2.4707	13.0	4.0
C/2007 VO53 Spacewatch	75.0008	59.7525	87.0201	7.0	4.0
P/2007 VQ11 Catalina	277.6735	164.0538	12.3251	12.0	4.0
C/2007 W1 Boattini	306.5504	334.5303	9.8892	9.5	4.0
C/2007 W3 LINEAR	112.6470	73.0641	78.6671	12.0	4.0
C/2007 Y1 LINEAR	357.1230	133.0925	110.1765	10.5	4.0
C/2007 Y2 McNaught	257.6879	303.4619	98.5030	9.0	4.0
C/2008 A1 McNaught	348.4818	277.8856	82.5507	6.5	4.0
P/2008 A2 LINEAR	233.1676	315.5218	19.1588	15.5	4.0
C/2008 C1 Chen-Gao	180.9290	307.7826	61.7844	10.0	4.0
C/2008 E1 Catalina	269.8707	189.0370	35.0385	7.0	4.0
C/2008 E3 Garradd	221.416	105.809	105.469	6.0	4.0
C/2008 G1 Gibbs	63.6956	215.9218	72.8687	9.5	4.0
C/2008 H1 LINEAR	96.084	34.616	75.497	10.5	4.0
C/2008 J1 Boattini	68.253	273.511	61.854	10.0	4.0
P/2008 J2 Beshore	143.165	97.432	10.622	9.0	4.0
C/2008 J3 McNaught	259.341	28.660	25.427	10.5	4.0
C/2008 J4 McNaught	93.709	289.109	88.396	15.0	4.0
C/2008 J5 (Garradd)	316.007	286.925	94.015	12.0	4.0
C/2008 J6 (Hill)	10.904	298.221	45.026	10.5	4.0
C/2008 L2 (Hill)	130.234	223.005	25.317	12.0	4.0
C/2008 L3 (Hill)	101.694	24.130	100.225	12.5	4.0
C/2008 N1 (Holmes)	100.746	357.472	115.532	9.0	4.0
P/2008 O2 (McNaught)	27.384	325.876	9.490	9.0	4.0
P/2008 O3 (Boattini)	341.249	47.609	32.266	13.0	4.0
C/2008 P1 (Garradd)	12.089	358.157	63.701	7.0	4.0
C/2008 Q1 (Maticic)	104.484	9.350	118.634	10.0	4.0

Name (nome)	w	Ω	I	G	H
P/2008 Q2 (Ory)	329.5913	60.7046	2.7550	16.5	4.0
C/2008 Q3 (Garradd)	340.8093	219.7199	140.7059	10.0	4.0
P/2008 QP20 (LINEAR-Hill)	71.9462	325.1471	7.7485	10.0	4.0
P/2008 R1 (Garradd)	256.5020	51.9968	15.9030	15.5	4.0
C/2008 R3 (LINEAR)	84.148	270.560	43.239	13.0	4.0
P/2008 S1 (Catalina-McNaught)	203.6141	111.3759	15.1077	15.0	4.0
C/2008 S3 (Boattini)	39.898	54.950	162.682	4.0	4.0
P/2008 T1 (Boattini)	36.0960	291.6690	2.0826	11.0	4.0
C/2008 T2 (Cardinal)	215.961	309.549	56.322	6.0	4.0
P/2008 T4 (Hill)	0.472	44.748	6.298	13.5	4.0
C/2008 X3 (LINEAR)	140.961	337.992	66.311	13.0	4.0
P/2008 Y1 (Boattini)	162.4578	259.7532	8.8036	15.0	4.0
P/2008 Y2 (Gibbs)	162.3501	330.8918	7.2752	16.0	4.0
P/2008 Y3 (McNaught)	238.2773	262.9372	38.8144	8.5	4.0
P/2009 B1 (Boattini)	128.6057	297.4434	22.2273	13.0	4.0
C/2009 B2 (LINEAR)	192.4647	18.8090	156.8693	12.5	4.0
C/2009 E1 (Itagaki)	47.130	105.780	128.398	11.5	4.0
C/2009 F1 (Larson)	219.081	357.926	171.397	15.0	4.0
C/2009 F2 (McNaught)	337.818	213.844	59.546	6.0	4.0
C/2009 F4 (McNaught)	260.763	53.542	79.148	3.0	4.0
C/2009 F5 (McNaught)	300.346	218.981	86.161	10.0	4.0
C/2009 F6 (Yi-SWAN)	129.799	278.704	85.748	6.0	4.0
C/2009 G1 (STEREO)	175.791	120.604	108.436	9.0	4.0
P/2009 K1 (Gibbs)	27.086	172.802	5.745	17.0	4.0
C/2009 K2 (Catalina)	147.770	123.771	66.828	10.0	4.0
C/2009 K3 (Beshore)	251.413	0.032	146.680	8.5	4.0
C/2009 K4 (Gibbs)	127.191	29.880	34.809	13.0	4.0
C/2009 K5 (McNaught)	66.229	257.879	103.856	7.5	4.0
P/2009 L2 (Yang-Gao)	346.918	259.318	16.219	15.0	4.0
C/2009 O2 Catalina	132.806	310.291	108.534	11.0	4.0
C/2009 O3 Hill	150.619	184.631	16.040	14.0	4.0
C/2009 O4 Hill	223.563	172.943	95.794	9.0	4.0
C/2009 P1 Garradd	90.6670	325.8665	106.3641	4.0	4.0
C/2009 P2 Boattini	76.0894	60.3921	163.4552	6.0	4.0
P/2009 Q1 Hill	157.9134	174.0866	14.4309	11.5	4.0
P/2009 Q4 Boattini	320.296	127.549	10.959	15.5	4.0
P/2009 Q5 McNaught	209.249	160.148	40.915	10.0	4.0
C/2009 R1 McNaught	130.8565	322.7529	76.6038	8.0	4.0
P/2009 S1 Gibbs	223.987	157.127	25.821	13.0	4.0
P/2009 S2 McNaught	230.3660	121.6146	28.4493	14.0	4.0
C/2009 S3 Lemmon	129.389	223.179	59.962	6.5	4.0
P/2009 SK280 Spacewatch-Hill	329.1322	36.3656	16.8657	11.0	4.0
C/2009 T1 McNaught	281.970	54.285	89.932	6.0	4.0
P/2009 T2 La Sagra	215.4594	215.9868	28.0991	14.0	4.0
C/2009 T3 LINEAR	32.407	60.095	148.742	12.5	4.0
C/2009 U1 Garradd	6.404	67.196	68.944	10.5	4.0
C/2009 U3 Hill	78.822	49.263	50.999	13.0	4.0
P/2009 U4 McNaught	261.923	54.361	10.218	14.0	4.0
P/2009 U5 Grauer	139.324	118.302	43.016	10.0	4.0
P/2009 U6 LINEAR	308.5827	112.4843	14.6376	13.0	4.0
P/2009 W1 Hill	49.150	55.223	14.523	11.5	4.0
C/2009 W2 Boattini	118.066	199.323	164.512	7.0	4.0
C/2009 Y1 Catalina	128.725	160.385	107.503	9.0	4.0
C/2009 Y2 Kowalski	177.247	260.726	30.655	13.0	4.0
P/2010 A1 Hill	13.259	47.356	10.327	13.0	4.0
P/2010 A2 LINEAR	132.714	320.272	5.259	15.5	4.0
P/2010 A3 Hill	41.2826	64.8299	15.0275	14.0	4.0
C/2010 A4 Siding Spring	271.518	346.730	96.788	10.5	4.0
P/2010 A5 LINEAR	308.100	277.284	5.762	13.0	4.0
C/2010 B1 Cardinal	211.5468	277.2086	101.9769	7.5	4.0
P/2010 B2 WISE	155.4007	0.6617	8.8897	17.0	4.0
P/2010 C1 Scotti	3.6353	142.0341	9.1426	9.5	4.0

Name (nome)	w	Ω	I	G	H
P/2010 D1 WISE	225.9262	160.9005	9.6471	13.0	4.0
P/2010 D2 WISE	119.034	319.834	57.181	11.0	4.0
C/2010 D3 WISE	304.6277	255.2317	76.4128	9.0	4.0
C/2010 D4 WISE	44.4626	266.8304	105.6598	6.5	4.0
C/2010 DG56 WISE	317.0731	4.0718	160.3957	16.0	4.0
C/2010 E1 Garradd	297.555	169.095	72.071	11.5	4.0
P/2010 E2 Jarnac	8.3191	177.8996	15.4412	14.0	4.0
C/2010 E3 WISE	49.974	117.328	96.480	15.0	4.0
C/2010 E5 Scotti	221.192	12.768	32.857	10.0	4.0
C/2010 F1 Boattini	127.4905	344.3971	64.9242	9.5	4.0
C/2010 F3 Scotti	31.1813	157.3940	4.6472	8.5	4.0
C/2010 F4 (Machholz)	120.718	237.294	89.143	13.5	4.0
C/2010 FB87 WISE-Garradd	265.0336	89.9026	107.6273	10.0	4.0
C/2010 G1 (Boattini)	168.618	287.430	78.384	12.5	4.0
C/2010 G2 Hill	137.4173	246.7826	103.7560	8.0	4.0
C/2010 G3 WISE	75.2100	313.7459	108.2500	8.5	4.0
C/2010 H1 (Garradd)	240.634	346.068	35.018	13.0	4.0
P/2010 H2 Vales	129.8170	64.3114	14.2572	6.0	4.0
P/2010 H4 (Scotti)	304.274	41.064	7.665	8.0	4.0
P/2010 H5 Scotti	174.6650	24.8965	14.0882	13.0	2.0
C/2010 J1 Boattini	333.117	254.815	134.386	12.0	4.0
C/2010 J2 McNaught	4.6594	311.8215	125.8564	9.0	4.0
P/2010 J3 McMillan	157.3575	106.6580	13.2559	11.0	4.0
C/2010 J4 WISE	83.751	316.404	162.297	19.5	4.0
P/2010 J5 McNaught	149.8739	65.6710	7.3564	10.0	4.0
P/2010 JC81 WISE	12.5823	30.7694	38.6900	9.0	4.0
P/2010 K2 WISE	328.584	281.251	10.616	19.0	4.0
C/2010 KW7 WISE	332.3889	104.8326	147.0845	15.0	2.0
P/2010 L1 WISE	317.105	318.609	21.113	16.0	4.0
C/2010 L3 Catalina	121.7515	38.2693	102.6356	4.5	4.0
C/2010 L4 WISE	103.939	124.519	103.711	11.5	4.0
C/2010 L5 WISE	215.9845	206.4301	146.9531	18.0	4.0
C/2010 M1 Gibbs	265.318	82.150	78.373	9.0	4.0
P/2010 N1 WISE	153.4939	113.2101	12.8764	17.0	4.0
P/2010 P4 WISE	354.4274	2.3361	24.1457	19.5	6.0
C/2010 R1 LINEAR	114.4997	343.6454	156.9341	6.0	4.0
P/2010 R2 La Sagra	59.0617	270.6494	21.4173	13.0	4.0
C/2010 S1 LINEAR	118.6115	93.4365	125.3352	3.5	4.0
P/2010 T1 McNaught	221.4397	130.0446	32.4933	11.0	4.0
P/2010 T2 PANSTARRS	356.1340	59.5943	8.0279	11.5	4.0
P/2010 TO20 LINEAR-Grauer	258.0117	44.1711	2.5821	9.0	4.0
P/2010 U1 Boattini	88.8513	281.2571	8.2263	9.5	4.0
P/2010 U2 Hill	44.2879	357.1526	16.8594	13.0	4.0
C/2010 U3 Boattini	87.9092	43.0310	55.4189	1.0	4.0
P/2010 UH55 Spacewatch	221.6578	235.2604	8.6624	11.0	4.0
P/2010 V1 Ikeya-Murakami	152.2462	3.7943	9.3797	8.0	4.0
C/2010 V1 Ikeya-Murakami	160.100	5.876	8.985	8.0	4.0
P/2010 WK LINEAR	40.8976	11.4932	11.4771	14.5	2.0
C/2010 X1 Elenin	343.8426	323.1697	1.8396	10.0	4.0
P/2011 A2 Scotti	95.2653	56.0511	4.5078	16.5	4.0
C/2011 A3 Gibbs	141.1623	124.8898	26.0760	10.0	4.0
C/2011 C1 McNaught	84.4200	192.4660	16.8326	15.5	4.0
P/2011 C2 Gibbs	160.5374	12.2043	10.9096	9.0	4.0
C/2011 C3 Gibbs	206.7822	20.9199	49.4013	17.0	4.0
C/2011 F1 LINEAR	192.5675	85.1185	56.6107	5.0	4.0
C/2011 G1 McNaught	354.5569	152.5920	162.2337	12.0	4.0
C/2011 J2 LINEAR	85.2458	163.9439	122.8096	6.0	4.0
C/2011 J3 LINEAR	27.7321	21.5272	114.7163	15.0	4.0
P/2011 JB15 Spacewatch-Boattin	110.8708	153.7535	19.1424	9.0	4.0
C/2011 K1 Schwartz-Holvorcem	166.9110	70.7531	122.4286	12.0	4.0
C/2011 L1 McNaught	294.4502	252.3666	65.5206	11.0	4.0
C/2011 L2 McNaught	257.0296	131.3505	104.2570	12.5	4.0

Name (nome)	w	Ω	I	G	H
C/2011 L3 McNaught	27.7309	307.7577	87.1144	12.5	4.0
C/2011 L4 PANSTARRS	333.6518	65.6658	84.2072	5.5	4.0
C/2011 M1 LINEAR	119.5845	324.7904	70.1797	12.0	4.0
P/2011 N1 ASH	331.0400	77.6838	35.6715	11.5	4.0
C/2011 N2 McNaught	357.0387	274.0198	33.6776	11.5	4.0
P/2011 NO1	263.9063	296.0719	15.1057	15.0	4.0
C/2011 O1 LINEAR	232.3842	89.8147	76.4961	7.0	4.0
P/2011 P1 McNaught	344.6909	7.4738	5.8393	9.0	4.0
C/2011 P2 PANSTARRS	76.3228	204.0097	8.9899	8.5	4.0
C/2011 Q1 PANSTARRS	136.0208	142.2495	94.8554	9.5	4.0
C/2011 Q2 McNaught	34.6288	287.3685	36.8670	10.0	4.0
P/2011 Q3 McNaught	310.2160	35.3177	6.0504	13.5	4.0
C/2011 Q4 SWAN	1.8751	252.0886	147.8435	12.5	4.0
C/2011 R1 McNaught	308.8573	221.4079	116.1968	6.5	4.0
P/2011 R2 PANSTARRS	230.6875	147.0822	4.9888	14.5	4.0
P/2011 R3 Novichonok-Gerke	225.1677	190.6211	19.2442	11.0	4.0
P/2011 S1 Gibbs	189.3352	218.1497	2.7157	9.5	4.0
C/2011 S1 Gibbs	174.6461	229.9225	2.2327	9.5	4.0
C/2011 S2 Kowalski	192.1959	288.0667	17.5724	15.5	4.0
P/2011 U1 PANSTARRS	353.1366	135.0078	15.2426	14.5	4.0
P/2011 U2 Bressi	157.6011	266.7739	9.6017	10.0	4.0
C/2011 U3 PANSTARRS	287.8395	228.3797	116.7800	14.0	4.0
P/2011 UA134 Spacewatch-PANSTA	32.3261	40.6362	10.5385	17.5	4.0
C/2011 UF305 LINEAR	121.9892	297.4350	93.9603	9.0	4.0
P/2011 V1 Boattini	259.6698	60.1425	7.3741	15.5	4.0
P/2011 VJ5 Lemmon	315.1180	169.9761	3.9719	17.5	4.0
P/2011 W1 PANSTARRS	282.6680	161.8839	3.7183	11.5	4.0
P/2011 W2 Rinner	221.0702	232.0174	13.7739	13.0	4.0
C/2011 W3 Lovejoy	53.6274	326.5407	134.4136	14.5	4.0
P/2011 Y1 Levy	179.6272	279.7450	18.2640	10.5	4.0
P/2011 Y2 Boattini	131.1924	310.0093	6.3517	15.0	4.0
C/2011 Y3 Boattini	341.3639	84.7844	26.3662	8.0	4.0
C/2012 A1 PANSTARRS	191.7066	277.9752	120.8965	6.0	4.0
C/2012 A2 LINEAR	101.5637	191.4172	125.8734	8.5	4.0
P/2012 B1 PANSTARRS	162.3908	36.0535	7.6257	9.0	4.0
C/2012 B3 La Sagra	49.2216	252.8441	106.8530	10.0	4.0
C/2012 C1 McNaught	284.5924	299.7748	95.6793	7.5	4.0
C/2012 C2 Bruenjes	62.9555	117.7558	162.7122	17.0	4.0
P/2012 C3 PANSTARRS	344.9542	135.3677	9.0920	13.0	4.0
C/2012 CH17 MOSS	138.0110	125.9743	27.7369	11.0	4.0
C/2012 E1 Hill	44.4964	286.1164	122.4703	6.5	4.0
C/2012 E3 PANSTARRS	106.9088	5.1767	106.3737	10.0	4.0
C/2012 F1 Gibbs	300.2191	129.7586	159.4309	14.0	4.0
P/2012 F2 PANSTARRS	38.5955	225.7736	15.1842	12.0	4.0
C/2012 F3 PANSTARRS	103.8393	164.6489	11.3018	6.5	4.0
P/2012 F5 Gibbs	175.8171	216.8063	9.7531	12.0	4.0
C/2012 F6 Lemmon	304.9873	332.7147	82.6078	10.0	4.0
P/2012 G1 PANSTARRS	286.1149	282.5440	11.6925	15.0	4.0
P/2012 H1 PANSTARRS	25.9071	126.3255	6.7436	13.0	4.0
C/2012 H2 McNaught	295.9177	184.2014	92.8362	15.5	4.0
C/2012 J1 Catalina	144.6291	235.3397	34.1723	8.0	4.0
C/2012 K1 PANSTARRS	199.5894	317.6772	142.3659	4.5	4.0
P/2012 K3 Gibbs	177.6246	124.8315	13.4004	15.0	4.0
C/2012 K5 LINEAR	138.8046	279.0618	92.8495	12.0	4.0
C/2012 K6 McNaught	338.4650	206.7851	135.2236	8.5	4.0
C/2012 K8 Lemmon	76.3671	312.7666	106.0480	6.0	4.0
C/2012 L1 LINEAR	140.4623	271.7259	87.2114	10.5	4.0
C/2012 L2 LINEAR	205.9492	270.2679	70.9859	10.0	4.0
C/2012 L3 LINEAR	106.8290	53.3787	134.2120	13.0	4.0
P/2012 NJ La Sagra	338.4140	315.7632	84.3755	14.5	4.0
P/2012 O1 McNaught	222.1212	111.2637	11.4244	17.5	4.0
P/2012 O2 McNaught	183.0486	120.8134	24.5272	17.0	4.0

Name (nome)	w	Ω	I	G	H
P/2012 O3 McNaught	343.6420	336.9982	16.4929	16.5	4.0
C/2012 Q1 Kowalski	144.6785	184.5166	45.0857	4.0	4.0
C/2012 S1 ISON	345.4991	295.7542	61.7662	6.0	4.0
P/2012 S2 La Sagra	312.1763	52.7263	8.6194	17.0	4.0
C/2012 S3 PANSTARRS	176.0429	119.1794	113.0162	10.0	4.0
C/2012 S4 PANSTARRS	165.8284	174.0762	126.0182	8.5	4.0
P/2012 SB6 Lemmon	12.6058	9.5541	10.9812	14.0	4.0
P/2012 T1 PANSTARRS	323.2225	83.9812	11.4000	15.0	4.0
P/2012 T2 PANSTARRS	253.7738	72.3013	12.7695	10.0	4.0
P/2012 T3 PANSTARRS	198.4921	113.8604	9.4895	15.0	4.0
C/2012 T4 McNaught	219.6306	99.3012	24.1074	13.5	4.0
C/2012 T5 Bressi	318.0972	230.5937	72.0944	13.0	4.0
C/2012 T6 Kowalski	196.4690	187.6624	34.2844	15.0	4.0
P/2012 T7 Vorobjov	174.7685	213.3434	13.5543	11.5	4.0
P/2012 TK8 Tenagra	128.0876	289.7266	6.2922	13.0	4.0
P/2012 TB36 Lemmon	36.4000	9.5860	6.9710	11.0	4.0
C/2012 U1 PANSTARRS	65.5897	28.0730	61.2306	7.5	4.0
P/2012 U2 PANSTARRS	213.6793	184.2301	9.9307	12.5	4.0
P/2012 US27 Siding Spring	1.2790	49.2127	39.2887	13.5	4.0
C/2012 V1 PANSTARRS	123.4550	85.3545	157.8359	12.0	4.0
C/2012 V2 LINEAR	217.3392	262.1725	67.1809	9.0	4.0
C/2012 X1 LINEAR	131.7698	113.3077	44.6647	8.0	4.0
C/2012 X2 PANSTARRS	216.1139	270.9300	34.0028	9.0	4.0
C/2012 Y1 LINEAR	245.1579	200.4168	22.5712	15.0	4.0
C/2012 Y3 McNaught	235.6696	122.7156	73.2271	11.0	4.0
C/2013 A1 Siding Spring	2.4677	300.9158	128.9960	6.0	4.0
P/2013 A2 Scotti	134.9585	355.7582	3.3732	15.5	4.0
P/2013 AL76 Catalina	27.2738	145.9074	144.8512	16.0	4.0
C/2013 B2 Catalina	156.1431	331.7963	43.5623	11.0	4.0
C/2013 C2 Tenagra	230.3256	249.8523	21.3262	10.0	4.0
P/2013 CE31 MOSS	27.1743	87.1220	4.7210	11.0	4.0
C/2013 D1 Holvorcem	250.2582	295.6486	10.2062	14.0	4.0
C/2013 E1 McNaught	304.9617	133.7322	158.7947	5.5	4.0
C/2013 E2 Iwamoto	95.0410	181.9233	21.9204	11.5	4.0
C/2013 F1 Boattini	96.2242	29.5904	80.5383	13.0	4.0
C/2013 F2 Catalina	122.9700	344.2716	61.7497	7.0	4.0
C/2013 F3 McNaught	18.7763	266.0995	85.4489	12.0	4.0
C/2013 G1 Kowalski	47.5715	221.3985	5.4154	11.0	4.0
C/2013 G2 McNaught	293.4372	274.1660	95.8469	11.5	4.0
C/2013 G3 PANSTARRS	77.0661	208.1174	63.7474	9.0	4.0
P/2013 G4 PANSTARRS	211.4021	340.1174	5.9571	15.0	4.0
C/2013 G5 Catalina	165.6003	144.3911	40.6694	14.0	4.0
C/2013 G6 Lemmon	215.5531	44.5671	124.1523	13.5	4.0
C/2013 G7 McNaught	217.9370	48.4014	105.1893	7.5	4.0
C/2013 G8 PANSTARRS	81.6681	241.1297	27.6471	8.5	4.0
C/2013 G9 Tenagra	210.2125	36.4007	145.6330	7.0	4.0
C/2013 H1 La Sagra	141.5131	84.2059	27.3289	13.0	4.0
C/2013 H2 Boattini	259.5192	255.1185	108.2292	6.5	4.0
P/2013 J2 McNaught	39.4115	288.3733	15.1360	13.0	4.0
C/2013 J3 McNaught	336.7797	201.2876	117.2606	10.0	4.0
P/2013 J4 PANSTARRS	81.5873	77.6293	4.7614	14.0	4.0

COMETE CON PASSAGGI VICINI ALLA TERRA NEC NEAR EARTH COMETS

(Elements are with respect to the J2000 heliocentric-ecliptic reference frame.)

e : Eccentricity of the orbit
q (AU) : Perihelion distance of the orbit in AU
MOID (AU) : Minimum orbit intersection distance (the minimum distance between the osculating orbits of the NEO and the Earth)

e : eccentricità dell'orbita
q : distanza perielica in U.A.
MOID : minima distanza dell'orbita della cometa alla Terra

Comet	q	e	MOID
1P/Halley	0.586000	0.967000	0.063782
2P/Encke	0.336127	0.848232	0.173100
3D/Biela	0.879000	0.751000	0.000518
4P/Faye	1.655402	0.568711	0.366559
7P/Pons-Winnecke	1.253563	0.634864	0.239198
8P/Tuttle	1.026929	0.819754	0.095310
12P/Pons-Brooks	0.774000	0.955000	0.187300
13P/Olbers	1.178000	0.930000	0.477199
15P/Finlay	0.968176	0.721732	0.013074
18D/Perrine-Mrkos	1.272000	0.643000	0.289230
20D/Westphal	1.254000	0.920000	0.468283
21P/Giacobini-Zinner	1.030484	0.707054	0.033183
23P/Brorsen-Metcalf	0.479000	0.972000	0.193872
24P/Schaumasse	1.213719	0.703483	0.282180
26P/Grigg-Skjellerup	1.086353	0.640143	0.080716
27P/Crommelin	0.747948	0.918822	0.215189
41P/Tuttle-Giacobini-Kresak	1.049404	0.660196	0.135384
45P/Honda-Mrkos-Pajdusakova	0.529704	0.824658	0.060071
46P/Wirtanen	1.052753	0.659174	0.071863
55P/Tempel-Tuttle	0.976427	0.905552	0.008481
66P/du Toit	1.274266	0.787701	0.430358
67P/Churyumov-Gerasimenko	1.263790	0.637770	0.255674
72D/Denning-Fujikawa	0.779727	0.819859	0.074729
73P/Schwassmann-Wachmann	0.942870	0.692248	0.045198
79P/du Toit-Hartley	1.123942	0.618599	0.133531
85P/Boethin	1.147535	0.775555	0.149899
96P/Machholz	0.123792	0.959181	0.333773
103P/Hartley 2	1.058690	0.695128	0.066822
109P/Swift-Tuttle	0.959516	0.963225	0.000892
122P/de Vico	0.659337	0.962708	0.316817
141P/Machholz	0.757705	0.749138	0.105492
161P/Hartley-IRAS	1.272193	0.835102	0.447117
162P/Siding Spring	1.235382	0.595697	0.234588
169P/NEAT	0.607608	0.766925	0.142236
177P/Barnard	1.107242	0.954404	0.310655
181P/Shoemaker-Levy	1.122634	0.707725	0.159639
182P/LONEOS	1.008530	0.659456	0.109366
185P/Petriew	0.931888	0.699431	0.063157
189P/NEAT	1.177232	0.596783	0.170302
197P/LINEAR	1.061639	0.629583	0.051758
206P/Barnard-Boattini	1.145196	0.646480	0.140964
207P/NEAT	0.944150	0.757147	0.147584
209P/LINEAR	0.913703	0.688980	0.039455
210P/Christensen	0.534907	0.831644	0.168080
217P/LINEAR	1.223968	0.689679	0.306437
222P/LINEAR	0.780162	0.727021	0.062411
225P/LINEAR	1.314741	0.639325	0.194983
249P/LINEAR	0.510880	0.816033	0.052555
252P/LINEAR	1.001368	0.672411	0.017389
255P/Levy	1.007477	0.668271	0.023582
262P/McNaught-Russell	1.279998	0.815338	0.291521
263P/Gibbs	1.251302	0.586980	0.284398
273P/Pons-Gambart	0.810196	0.975320	0.171685

Comet	q	e	MOID
D/1766 G1 (Helfenzrieder)	0.848	0.406	n/a
D/1770 L1 (Lexell)	0.786	0.674	n/a
D/1819 W1 (Blanpain)	0.699	0.892	n/a
D/1884 O1 (Barnard)	0.583	1.279	n/a
D/1894 F1 (Denning)	0.698	1.147	n/a
D/1895 Q1 (Swift)	0.652	1.298	n/a
C/1917 F1 (Mellish)	0.993	0.190	n/a
C/1921 H1 (Dubiago)	0.860	1.101	0.369848
C/1937 D1 (Wilk)	0.981	0.619	n/a
C/1942 EA (Vaisala)	0.934	1.287	n/a
D/1978 R1 (Haneda-Campos)	0.665	1.101	0.134915
C/1989 A3 (Bradfield)	0.978	0.420	n/a
C/1991 L3 (Levy)	0.929	0.983	0.075241
P/1999 J6 (SOHO)	0.984	0.049	0.010175
P/1999 R1 (SOHO)	0.977	0.057	0.076579
P/1999 RO28 (LONEOS)	0.651	1.232	0.247638
C/1999 X3 (SOHO)	0.982	0.048	0.031226
C/2001 OG108 (LONEOS)	0.925	0.994	0.300836
C/2001 W2 (BATTERS)	0.941	1.051	0.145020
C/2002 R5 (SOHO)	0.985	0.047	0.094044
P/2002 S7 (SOHO)	0.985	0.048	0.114346
P/2003 O3 (LINEAR)	0.599	1.246	0.237830
P/2003 T12 (SOHO)	0.776	0.575	0.154911
P/2004 R1 (McNaught)	0.682	0.988	0.026001
P/2005 JQ5 (Catalina)	0.693	0.826	0.024400
P/2005 T4 (SWAN)	0.930	0.649	0.157276
P/2005 W4 (SOHO)	0.982	0.054	0.107655
P/2006 HR30 (Siding Spring)	0.843	1.226	0.452907
P/2007 T2 (Kowalski)	0.775	0.696	0.156287
P/2008 S1 (McNaught)	0.666	1.190	0.193761
P/2008 Y1 (Boattini)	0.735	1.272	0.289353
P/2009 L2 (Yang-Gao)	0.621	1.296	0.287468
P/2009 WX51 (Catalina)	0.740	0.800	0.009064
P/2010 K2 (WISE)	0.589	1.198	0.203811
C/2010 L5 (WISE)	0.904	0.791	0.114134
P/2011 NO1 (Elenin)	0.777	1.243	0.380673
C/2011 S2 (Kowalski)	0.932	1.115	0.137895
P/2012 NJ (La Sagra)	0.848	1.292	0.315315

PERIELII - PERIHELIA

Date dei perielii e distanza q dal Sole
Dates of the perihelium and distance q

Cometa-comet	Date	A.U.
C/2011 F1 LINEAR	08/01/2013	1,81877
C/2011 S1 Gibbs	16/01/2013	4,98785
276P/Vorobjov	18/01/2013	3,92276
259P/Garradd	24/01/2013	1,79307
P/2008 R1 (Garradd)	24/01/2013	1,79306
246P/NEAT	26/01/2013	2,8678
111P/Helin-Roman-Crockett	30/01/2013	3,7043
P/2013 G4 PANSTARRS	31/01/2013	2,60366
P/2012 US27 Siding Spring	08/02/2013	1,82065
P/2013 A2 Scotti	08/02/2013	2,17925
125P/Spacewatch	17/02/2013	1,52546
120P/Mueller 1	22/02/2013	2,72904
274P/Tombaugh-Tenagra	23/02/2013	2,44178
C/2012 T5 Bressi	24/02/2013	0,32281
P/2007 T2 Kowalski	25/02/2013	0,69593
91P/Russell 3	01/03/2013	2,61655
C/2012 C1 McNaught	01/03/2013	4,62967
C/2013 E1 McNaught	07/03/2013	7,80978
C/2013 E2 Iwamoto	08/03/2013	1,39233
272P/NEAT	10/03/2013	2,4396
C/2011 L4 PANSTARRS	10/03/2013	0,30154
256P/LINEAR	16/03/2013	2,68038
197P/LINEAR	24/03/2013	1,06146
C/2012 F6 Lemmon	24/03/2013	0,73125
C/2012 X2 PANSTARRS	03/04/2013	4,73537
P/2012 F2 PANSTARRS	07/04/2013	2,73199
C/2013 D1 Holvorcem	08/04/2013	2,48515
63P/Wild 1	10/04/2013	1,95049
C/2013 F2 Catalina	18/04/2013	6,21783
76P/West-Kohoutek-Ikemura	07/05/2013	1,60024
C/2012 L2 LINEAR	08/05/2013	1,49859
P/2012 TK8 Tenagra	09/05/2013	3,0897
114P/Wiseman-Skiff	13/05/2013	1,57484
C/2010 S1 LINEAR	20/05/2013	5,90022
C/2012 K6 McNaught	21/05/2013	3,36632
P/2010 A2 LINEAR	22/05/2013	2,00593
C/2013 J3 McNaught	23/05/2013	4,03817
C/2013 F3 McNaught	25/05/2013	2,25182
257P/Catalina	01/06/2013	2,12595
C/2013 H1 La Sagra	01/06/2013	2,62222
P/2005 JY126 (Catalina)	01/06/2013	2,12588
175P/Hergenrother	05/06/2013	2,05472
277P/LINEAR	11/06/2013	1,91907
112P/Urata-Niijima	29/06/2013	1,46468
C/2013 B2 Catalina	30/06/2013	3,75332
26P/Grigg-Skjellerup	06/07/2013	1,08589
46P/Wirtanen	12/07/2013	1,05741
C/2012 S4 PANSTARRS	12/07/2013	4,29418
P/2012 TB36 Lemmon	18/07/2013	4,26633
271P/van Houten-Lemmon	20/07/2013	4,25676
P/2012 B1 PANSTARRS	20/07/2013	3,80934
C/2012 V1 PANSTARRS	21/07/2013	2,08559
270P/Gehrels	23/07/2013	3,59778
C/2013 G6 Lemmon	24/07/2013	2,05047

Cometa-comet	Date	A.U.
P/2013 J4 PANSTARRS	24/07/2013	2,28767
84P/Giclas	26/07/2013	1,85165
178P/Hug-Bell	28/07/2013	1,94694
278P/McNaught	30/07/2013	2,0867
98P/Takamizawa	05/08/2013	1,67346
C/2012 V2 LINEAR	16/08/2013	1,4546
79P/du Toit-Hartley	23/08/2013	1,1239
P/2013 J2 McNaught	27/08/2013	2,08685
102P/Shoemaker 1	30/08/2013	1,97383
C/2013 G5 Catalina	01/09/2013	0,92916
C/2012 S3 PANSTARRS	03/09/2013	2,5988
266P/Christensen	04/09/2013	2,32578
P/2006 U5 (Christensen)	06/09/2013	2,3259
P/2007 C1 (Christensen)	11/09/2013	2,05079
121P/Shoemaker-Holt 2	12/09/2013	3,75305
184P/Lovas 2	26/09/2013	1,45201
154P/Brewington	17/10/2013	1,59037
P/2005 L1 (McNaught)	16/11/2013	3,14402
2P/Encke	21/11/2013	0,33625
C/2013 G1 Kowalski	24/11/2013	3,4042
C/2013 G8 PANSTARRS	27/11/2013	5,11124
C/2012 A1 PANSTARRS	29/11/2013	7,60636
C/2012 S1 ISON	29/11/2013	0,05301
25D/Neujmin 2	30/11/2013	1,33817
280P/Larsen	07/12/2013	2,62041
C/2011 J2 LINEAR	25/12/2013	3,44522
87P/Bus	07/01/2014	2,17342
P/2006 XG16 (Spacewatch)	10/01/2014	2,10221
P/2007 R2 Gibbs	12/01/2014	1,46589
32P/Comas Sola	20/01/2014	1,83366
129P/Shoemaker-Levy 3	06/02/2014	3,91341
169P/NEAT	12/02/2014	0,60653
C/2012 X1 LINEAR	16/02/2014	1,58017
52P/Harrington-Abell	26/02/2014	1,75711
P/2008 A2 LINEAR	26/02/2014	1,30541
P/2007 H3 (Garradd)	03/03/2014	1,82927
C/2013 G7 McNaught	16/03/2014	4,6907
117P/Helin-Roman-Alu 1	27/03/2014	3,05576
17P/Holmes	27/03/2014	2,05635
119P/Parker-Hartley	02/04/2014	3,02696
124P/Mrkos	09/04/2014	1,64488
156P/Russell-LINEAR	17/04/2014	1,5931
209P/LINEAR	28/04/2014	0,91244
191P/McNaught	02/05/2014	2,04791
134P/Kowal-Vavrova	21/05/2014	2,57234
132P/Helin-Roman-Alu 2	29/05/2014	1,92416
4P/Faye	02/06/2014	1,66684
16P/Brooks 2	04/06/2014	1,46657
181P/Shoemaker-Levy 6	06/06/2014	1,12749
222P/LINEAR	01/07/2014	0,78105
75D/Kohoutek	03/07/2014	1,78466
C/2012 U1 PANSTARRS	17/07/2014	5,69092
106P/Schuster	23/07/2014	1,55612
193P/LINEAR-NEAT	27/07/2014	2,04444

Cometa-comet	Date	A.U.
P/2011 S1 Gibbs	07/08/2014	7,19243
P/2008 J2 Beshore	10/08/2014	2,45485
206P/Barnard-Boattini	17/08/2014	1,13648
P/2008 Q2 (Ory)	19/08/2014	1,38223
C/2012 K8 Lemmon	23/08/2014	6,4261
C/2007 E1 (Garradd)	31/08/2014	18,8462
210P/Christensen	06/09/2014	0,54642
P/2007 H1 (McNaught)	07/09/2014	2,28231
170P/Christensen	15/09/2014	2,92844
C/2012 K1 PANSTARRS	27/09/2014	1,24656
11P/Tempel-Swift-LINEAR	29/09/2014	1,58407
72D/Denning-Fujikawa	10/10/2014	0,77973
108P/Ciffreo	21/10/2014	1,71912
70P/Kojima	25/10/2014	2,01189
C/2013 A1 Siding Spring	25/10/2014	1,39579
269P/Jedicke	06/11/2014	4,04813
80P/Peters-Hartley	08/11/2014	1,62392
C/2013 G3 PANSTARRS	10/11/2014	3,79646
40P/Vaisala 1	18/11/2014	1,79597
135P/Shoemaker-Levy 8	07/12/2014	2,7211
110P/Hartley 3	19/12/2014	2,48439
P/2006 R2 (Christensen)	19/12/2014	3,03924
15P/Finlay	22/12/2014	0,97001
201P/LONEOS	26/01/2015	1,35801
C/2013 G9 Tenagra	30/01/2015	5,03679
7P/Pons-Winnecke	05/02/2015	1,2533
P/2010 B2 WISE	24/02/2015	1,60697
6P/d'Arrest	26/02/2015	1,3535
92P/Sanguin	27/02/2015	1,80804
268P/Bernardi	01/03/2015	2,34535
44P/Reinmuth 2	24/03/2015	2,1162
42P/Neujmin 3	30/03/2015	2,01471
C/2012 F3 PANSTARRS	05/04/2015	3,46833
88P/Howell	06/04/2015	1,36157
86P/Wild 3	19/04/2015	2,30113
174P/Echeclus	22/04/2015	5,81566
113P/Spitaler	25/04/2015	2,12821
P/2007 S1 Zhao	02/05/2015	2,49438
P/2009 Q4 Boattini	04/05/2015	1,31873
205P/Giacobini	09/05/2015	1,52638
P/2008 QP20 (LINEAR-Hill)	11/05/2015	1,72331
172P/Yeung	14/05/2015	2,24099
57P/duToit-Neujmin-Delporte	22/05/2015	1,72471
19P/Borrelly	29/05/2015	1,35365
34D/Gale	29/05/2015	1,18291
P/2010 K2 WISE	29/05/2015	1,19683
P/2012 F5 Gibbs	02/06/2015	2,87264
220P/McNaught	17/06/2015	1,55397
148P/Anderson-LINEAR	18/06/2015	1,70266
196P/Tichy	21/06/2015	2,1516
233P/La Sagra	27/06/2015	1,79498
162P/Siding Spring	04/07/2015	1,22767
P/2008 S1 (Catalina-McNaught)	08/07/2015	1,19052
140P/Bowell-Skiff	19/07/2015	1,97188

Cometa-comet	Date	A.U.
P/2004 FY140 (LINEAR)	24/07/2015	4,1076
218P/LINEAR	30/07/2015	1,70272
221P/LINEAR	30/07/2015	1,79027
51P/Harrington	05/08/2015	1,68794
141P/Machholz 2-D	06/08/2015	0,74901
67P/Churyumov-Gerasimenko	12/08/2015	1,24288
C/2011 Q4 SWAN	12/08/2015	12,214
141P/Machholz 2-A	15/08/2015	0,75285
61P/Shajn-Schaldach	27/09/2015	2,10838
5D/Brorsen	11/10/2015	0,58985
P/2007 V2 Hill	14/10/2015	2,77478
P/2009 L2 (Yang-Gao)	23/10/2015	1,29714
22P/Kopff	27/10/2015	1,57258
P/2009 U6 LINEAR	28/10/2015	1,4842
151P/Helin	01/11/2015	2,53184
18D/Perrine-Mrkos	06/11/2015	1,27225
214P/LINEAR	06/11/2015	1,83984
P/2005 RV25 (LONEOS-Christensen)	08/11/2015	3,60784
P/2008 Y2 (Gibbs)	11/11/2015	1,63839
230P/LINEAR	14/11/2015	1,486
10P/Tempel 2	15/11/2015	1,42103
249P/LINEAR	30/11/2015	0,51059
P/2010 R2 La Sagra	03/12/2015	2,6192
224P/LINEAR-NEAT	06/12/2015	1,88186
180P/NEAT	10/12/2015	2,46868
204P/LINEAR-NEAT	11/12/2015	1,93852
P/1998 QP54 (LONEOS-Tucker)	15/12/2015	1,87992
83D/Russell 1	01/01/2016	1,61154
116P/Wild 4	11/01/2016	2,17519
211P/Hill	31/01/2016	2,36164
50P/Arend	06/02/2016	1,92373
225P/LINEAR	13/02/2016	1,19206
147P/Kushida-Muramatsu	27/02/2016	2,75623
P/2010 V1 Ikeya-Murakami	28/02/2016	1,5765
194P/LINEAR	09/03/2016	1,70702
127P/Holt-Olmstead	14/03/2016	2,19355
252P/LINEAR	22/03/2016	1,00137
100P/Hartley 1	30/03/2016	1,99088
190P/Mueller	30/03/2016	2,03575
73P/Schwassmann-Wachmann 3-AZ	11/04/2016	0,94032
73P/Schwassmann-Wachmann 3-AY	12/04/2016	0,94019
53P/Van Biesbroeck	19/04/2016	2,41487
77P/Longmore	05/05/2016	2,31065
73P/Schwassmann-Wachmann 3-AF	06/05/2016	0,94118
P/2010 N1 WISE	12/05/2016	1,49452
73P/Schwassmann-Wachmann 3-AW	19/05/2016	0,94026
P/2007 R3 Gibbs	21/05/2016	2,50361
136P/Mueller 3	25/05/2016	2,96302
216P/LINEAR	30/05/2016	2,1526
202P/Scotti	06/06/2016	2,52548
157P/Tritton	12/06/2016	1,35798
118P/Shoemaker-Levy 4	16/06/2016	1,98165
146P/Shoemaker-LINEAR	19/06/2016	1,41774
208P/McMillan	21/06/2016	2,52521

Cometa-comet	Date	A.U.
207P/NEAT	03/07/2016	0,944
P/2009 K1 (Gibbs)	05/07/2016	1,32261
P/2004 VR8 (LONEOS)	08/07/2016	2,37653
73P/Schwassmann-Wachmann 3-AA	10/07/2016	0,93928
279P/La Sagra	15/07/2016	2,14771
167P/CINEOS	19/07/2016	1,78354
81P/Wild 2	21/07/2016	1,59618
9P/Tempel 1	21/07/2016	1,50908
150P/LONEOS	28/07/2016	1,76286
73P/Schwassmann-Wachmann 3-AJ	28/07/2016	0,93608
56P/Slaughter-Burnham	31/07/2016	2,53497
73P/Schwassmann-Wachmann 3-BE	09/08/2016	0,93942
73P/Schwassmann-Wachmann 3-BH	13/08/2016	0,93982
73P/Schwassmann-Wachmann 3-BM	13/08/2016	0,93976
73P/Schwassmann-Wachmann 3-BP	13/08/2016	0,93978
73P/Schwassmann-Wachmann 3-BA	14/08/2016	0,94025
73P/Schwassmann-Wachmann 3-BI	14/08/2016	0,93977
43P/Wolf-Harrington	17/08/2016	1,35756
33P/Daniel	24/08/2016	2,16952
73P/Schwassmann-Wachmann 3-BB	27/08/2016	0,94242
144P/Kushida	03/09/2016	1,43901
212P/NEAT	14/09/2016	1,65423
73P/Schwassmann-Wachmann 3-V	14/09/2016	0,91221
104P/Kowal 2	15/09/2016	1,3961
73P/Schwassmann-Wachmann 3-BD	18/09/2016	0,94616
73P/Schwassmann-Wachmann 3-BG	21/09/2016	0,93934
73P/Schwassmann-Wachmann 3-BR	27/09/2016	0,9396
73P/Schwassmann-Wachmann 3-BF	07/10/2016	0,93957
94P/Russell 4	28/10/2016	2,23565
73P/Schwassmann-Wachmann 3-AT	29/10/2016	0,93963
238P/Read	30/10/2016	2,36467
P/2010 A2 LINEAR	09/11/2016	2,00593
P/2008 T1 (Boattini)	10/11/2016	3,04484
128P/Shoemaker-Holt 1-A	26/11/2016	3,04728
P/2009 S1 Gibbs	09/12/2016	2,41458
73P/Schwassmann-Wachmann 3-U	15/12/2016	0,93939
45P/Honda-Mrkos-Pajdusakova	29/12/2016	0,52976
73P/Schwassmann-Wachmann 3-AM	05/01/2017	0,93094
89P/Russell 2	07/01/2017	2,27994
128P/Shoemaker-Holt 1-B	13/01/2017	3,06883
73P/Schwassmann-Wachmann 3-BJ	14/01/2017	0,93925
73P/Schwassmann-Wachmann 3-AB	17/01/2017	0,93917
73P/Schwassmann-Wachmann 3-J	21/01/2017	0,93902
73P/Schwassmann-Wachmann 3-BK	23/01/2017	0,93927
73P/Schwassmann-Wachmann 3-BN	26/01/2017	0,93918
73P/Schwassmann-Wachmann 3-BO	28/01/2017	0,93922
188P/LINEAR-Mueller	31/01/2017	2,55256
73P/Schwassmann-Wachmann 3-AX	02/02/2017	0,93801
73P/Schwassmann-Wachmann 3	06/02/2017	0,93277
73P/Schwassmann-Wachmann 3-N	06/02/2017	0,93925
73P/Schwassmann-Wachmann 3-BQ	08/02/2017	0,93915
73P/Schwassmann-Wachmann 3-AK	16/02/2017	0,93771
73P/Schwassmann-Wachmann 3-AU	16/02/2017	0,9393
182P/LONEOS	18/02/2017	0,97802

Cometa-comet	Date	A.U.
237P/LINEAR	21/02/2017	2,41934
73P/Schwassmann-Wachmann 3-AV	23/02/2017	0,93902
73P/Schwassmann-Wachmann 3-B	23/02/2017	0,93911
219P/LINEAR	25/02/2017	2,36394
73P/Schwassmann-Wachmann 3-C	25/02/2017	0,94286
73P/Schwassmann-Wachmann 3-E	25/02/2017	0,93744
73P/Schwassmann-Wachmann 3-G	26/02/2017	0,93918
73P/Schwassmann-Wachmann 3-K	28/02/2017	0,93908
93P/Lovas 1	28/02/2017	1,7041
73P/Schwassmann-Wachmann 3-H	01/03/2017	0,93917
73P/Schwassmann-Wachmann 3-AD	07/03/2017	0,94071
73P/Schwassmann-Wachmann 3-M	08/03/2017	0,9391
2P/Encke	09/03/2017	0,33625
P/2006 G1 (McNaught)	10/03/2017	2,63209
73P/Schwassmann-Wachmann 3-R	11/03/2017	0,93913
176P/LINEAR	15/03/2017	2,57553
73P/Schwassmann-Wachmann 3-AO	15/03/2017	0,94075
73P/Schwassmann-Wachmann 3-Z	15/03/2017	0,94204
73P/Schwassmann-Wachmann 3-AH	22/03/2017	0,94299
73P/Schwassmann-Wachmann 3-AN	23/03/2017	0,939
73P/Schwassmann-Wachmann 3-L	23/03/2017	0,93909
73P/Schwassmann-Wachmann 3-X	24/03/2017	0,93908
P/2006 T1 (Levy)	02/04/2017	0,9896
226P/Pigott-LINEAR-Kowalski	04/04/2017	1,91685
73P/Schwassmann-Wachmann 3-AC	05/04/2017	0,93905
73P/Schwassmann-Wachmann 3-P	09/04/2017	0,93916
73P/Schwassmann-Wachmann 3-AR	12/04/2017	0,9392
41P/Tuttle-Giacobini-Kresak	13/04/2017	1,04778
73P/Schwassmann-Wachmann 3-AG	13/04/2017	0,93904
73P/Schwassmann-Wachmann 3-AQ	14/04/2017	0,93901
54P/de Vico-Swift-NEAT	16/04/2017	2,17176
103P/Hartley 2	18/04/2017	1,05869
73P/Schwassmann-Wachmann 3-BL	22/04/2017	0,9391
90P/Gehrels 1	30/04/2017	2,96591
255P/Levy	01/05/2017	1,00758
73P/Schwassmann-Wachmann 3-AE	01/05/2017	0,93905
P/2011 Y1 Levy	01/05/2017	1,00745
229P/Gibbs	14/05/2017	2,44021
73P/Schwassmann-Wachmann 3-W	18/05/2017	0,93913
73P/Schwassmann-Wachmann 3-AI	26/05/2017	0,9389
73P/Schwassmann-Wachmann 3-BS	04/06/2017	0,93895
P/1999 XN120 (Catalina)	04/06/2017	3,30298
234P/LINEAR	09/06/2017	2,85633
47P/Ashbrook-Jackson	10/06/2017	2,80981
73P/Schwassmann-Wachmann 3-Q	12/06/2017	0,93971
227P/Catalina-LINEAR	21/06/2017	1,79592
71P/Clark	26/06/2017	1,56927
65P/Gunn	01/07/2017	2,62535
73P/Schwassmann-Wachmann 3-AL	01/07/2017	0,93877
251P/LINEAR	02/07/2017	1,7115
217P/LINEAR	07/07/2017	1,22348
73P/Schwassmann-Wachmann 3-BC	12/07/2017	0,93881
263P/Gibbs	16/07/2017	1,2513
189P/NEAT	17/07/2017	1,17756

Cometa-comet	Date	A.U.
73P/Schwassmann-Wachmann 3-AS	24/07/2017	0,93882
259P/Garradd	25/07/2017	1,79307
P/2008 R1 (Garradd)	27/07/2017	1,79306
73P/Schwassmann-Wachmann 3-AP	31/07/2017	0,93875
C/2010 F1 Boattini	01/08/2017	18,03992
30P/Reinmuth 1	19/08/2017	1,88319
P/2007 T6 Catalina	05/09/2017	2,23777
P/2010 H2 Vales	21/09/2017	3,10566
96P/Machholz 1	28/09/2017	0,25245
213P/Van Ness	19/10/2017	2,12238
24P/Schaumasse	30/10/2017	1,20501
236P/LINEAR	19/11/2017	1,83108
14P/Wolf	24/11/2017	2,7248
139P/Vaisala-Oterma	27/11/2017	3,40275
P/2011 A2 Scotti	29/11/2017	1,58639
183P/Korlevic-Juric	06/12/2017	3,89437
P/2010 D1 WISE	06/12/2017	2,66908
145P/Shoemaker-Levy 5	14/12/2017	1,987
P/2009 S2 McNaught	19/12/2017	2,2036
P/1998 VS24 (LINEAR)	26/12/2017	3,42241
P/2010 P4 WISE	11/01/2018	1,86247
74P/Smirnova-Chernykh	25/01/2018	3,54877
250P/Larson	27/01/2018	2,21403
185P/Petriew	28/01/2018	0,93184
197P/LINEAR	30/01/2018	1,06146
245P/WISE	17/02/2018	2,14099
P/2010 J5 McNaught	25/02/2018	3,74762
130P/McNaught-Hughes	26/02/2018	2,08467
187P/LINEAR	27/02/2018	3,69532
P/2011 VJ5 Lemmon	03/03/2018	1,50578
62P/Tsuchinshan 1	12/03/2018	1,48905
73P/Schwassmann-Wachmann 3-T	19/03/2018	0,93932
P/2012 T1 PANSTARRS	19/03/2018	2,41916
235P/LINEAR	21/03/2018	2,74091
37P/Forbes	23/04/2018	1,57822
P/2006 F1 (Kowalski)	23/04/2018	4,11995
P/2008 T4 (Hill)	24/04/2018	2,51086
169P/NEAT	27/04/2018	0,60653
253P/PANSTARRS	11/05/2018	2,03802
66P/du Toit	12/05/2018	1,27427
240P/NEAT	13/05/2018	2,12638
143P/Kowal-Mrkos	16/05/2018	2,53815
164P/Christensen	22/05/2018	1,67361
P/2011 R2 PANSTARRS	31/05/2018	2,04923
P/2010 L1 WISE	01/06/2018	2,1489
82P/Gehrels 3	18/06/2018	3,63331
159P/LONEOS	27/06/2018	3,65084
215P/NEAT	02/07/2018	3,21596
49P/Arend-Rigaux	12/07/2018	1,42434
P/2007 T2 Kowalski	03/08/2018	0,69593
267P/LONEOS	09/08/2018	1,33776
73P/Schwassmann-Wachmann 3-S	12/08/2018	1,02237
105P/Singer Brewster	14/08/2018	2,05052
38P/Stephan-Oterma	26/08/2018	1,57442

Cometa-comet	Date	A.U.
125P/Spacewatch	30/08/2018	1,52546
243P/NEAT	06/09/2018	2,45754
73P/Schwassmann-Wachmann 3-Y	09/09/2018	0,93916
59P/Kearns-Kwee	11/09/2018	2,35521
79P/du Toit-Hartley	13/09/2018	1,1239
P/2010 A1 Hill	14/09/2018	1,95063
21P/Giacobini-Zinner	15/09/2018	1,03051
48P/Johnson	21/09/2018	2,29998
26P/Grigg-Skjellerup	02/10/2018	1,08589
64P/Swift-Gehrels	20/10/2018	1,37711
P/2011 V1 Boattini	25/10/2018	1,71167
P/2009 W1 Hill	26/10/2018	2,93817
186P/Garradd	01/11/2018	4,26391
P/2008 O2 (McNaught)	13/11/2018	3,79833
198P/ODAS	25/11/2018	1,97819
137P/Shoemaker-Levy 2	29/11/2018	1,90606
60P/Tsuchinshan 2	06/12/2018	1,61816
46P/Wirtanen	19/12/2018	1,05741
3D/Biela	25/12/2018	0,87907
247P/LINEAR	26/12/2018	1,49615
239P/LINEAR	30/12/2018	1,64506
P/2007 V1 Larson	04/01/2019	2,67658
P/2007 T4 Gibbs	08/01/2019	1,99968
171P/Spahr	09/01/2019	1,76498
223P/Skiff	16/01/2019	2,41058
131P/Mueller 2	28/01/2019	2,41675
123P/West-Hartley	03/02/2019	2,12871
C/2010 U3 Boattini	26/02/2019	8,46798
149P/Mueller 4	27/02/2019	2,6509
69P/Taylor	10/03/2019	2,27285
232P/Hill	30/03/2019	2,98319
78P/Gehrels 2	02/04/2019	2,0085
P/2011 W2 Rinner	02/04/2019	2,30311
P/2012 O1 McNaught	08/04/2019	1,4989
29P/Schwassmann-Wachmann 1	14/04/2019	5,74476
P/2012 O2 McNaught	21/04/2019	1,66083
222P/LINEAR	01/05/2019	0,78105
138P/Shoemaker-Levy 7	04/05/2019	1,70034
25D/Neujmin 2	07/05/2019	1,33817
209P/LINEAR	11/05/2019	0,91244
231P/LINEAR-NEAT	13/06/2019	3,03121
31P/Schwassmann-Wachmann 2	01/07/2019	3,42307
P/2008 Y1 (Boattini)	05/07/2019	1,27061
261P/Larson	16/07/2019	2,18969
264P/Larsen	22/07/2019	2,43729
200P/Larsen	24/07/2019	3,28038
163P/NEAT	31/07/2019	2,05709
168P/Hergenrother	26/08/2019	1,41484
P/2010 U2 Hill	10/09/2019	2,55374
260P/McNaught	10/10/2019	1,4979
76P/West-Kohoutek-Ikemura	25/10/2019	1,60024
P/2009 SK280 Spacewatch-Hill	31/10/2019	4,20758
P/2008 A2 LINEAR	12/11/2019	1,30541
68P/Klemola	21/11/2019	1,75902

Cometa-comet	Date	A.U.
P/2012 K3 Gibbs	03/12/2019	2,03149
101P/Chernykh	04/12/2019	2,35062
175P/Hergenrother	31/12/2019	2,05472
114P/Wiseman-Skiff	14/01/2020	1,57484
155P/Shoemaker 3	16/01/2020	1,81334
203P/Korlevic	11/02/2020	3,1824
228P/LINEAR	21/02/2020	3,42808
112P/Urata-Niijima	28/02/2020	1,46469
P/2010 A5 LINEAR	17/03/2020	1,69746
P/2007 C1 (Christensen)	20/03/2020	2,05079
P/2010 K2 WISE	19/04/2020	1,19683
266P/Christensen	22/04/2020	2,32578
124P/Mrkos	23/04/2020	1,64488
P/2006 U5 (Christensen)	24/04/2020	2,3259
P/2010 A2 LINEAR	28/04/2020	2,00593
P/2010 B2 WISE	30/04/2020	1,60697
167P/CINEOS	13/05/2020	1,78354
210P/Christensen	22/05/2020	0,54642
P/2007 R2 Gibbs	01/06/2020	1,46589
258P/PANSTARRS	07/06/2020	3,4792
206P/Barnard-Boattini	10/06/2020	1,13648
P/2012 H1 PANSTARRS	12/06/2020	3,47997
P/2008 Q2 (Ory)	19/06/2020	1,38223
2P/Encke	24/06/2020	0,33625
184P/Lovas 2	28/06/2020	1,45201
58P/Jackson-Neujmin	28/06/2020	1,38117
36P/Whipple	06/07/2020	3,08188
87P/Bus	10/07/2020	2,17342
84P/Giclas	14/07/2020	1,85166
249P/LINEAR	16/07/2020	0,51059
115P/Maury	22/07/2020	2,0397
P/2012 SB6 Lemmon	24/07/2020	2,40608
16P/Brooks 2	26/07/2020	1,46657
P/2011 U1 PANSTARRS	11/08/2020	2,35619
P/2012 F5 Gibbs	15/08/2020	2,87264
178P/Hug-Bell	18/08/2020	1,94694
160P/LINEAR	20/08/2020	2,07021
278P/McNaught	01/09/2020	2,0867
257P/Catalina	08/09/2020	2,12595
P/2005 JY126 (Catalina)	08/09/2020	2,12588
P/2007 VQ11 Catalina	14/09/2020	2,69363
P/2007 H3 (Garradd)	18/09/2020	1,82927
88P/Howell	29/09/2020	1,36157
85P/Boethin	03/10/2020	1,13466
233P/La Sagra	12/10/2020	1,79498
P/2009 Q4 Boattini	15/10/2020	1,31873
P/2005 Y2 (McNaught)	18/10/2020	3,35576
141P/Machholz 2-D	25/10/2020	0,749
162P/Siding Spring	28/10/2020	1,22767
141P/Machholz 2-A	06/11/2020	0,75285
91P/Russell 3	11/11/2020	2,61655
P/2007 Q2 Gilmore	15/11/2020	1,83903
102P/Shoemaker 1	22/11/2020	1,97384
P/2008 J2 Beshore	24/11/2020	2,45485

Cometa-comet	Date	A.U.
254P/McNaught	30/11/2020	3,21358
P/2012 G1 PANSTARRS	09/12/2020	2,58417
P/2006 XG16 (Spacewatch)	11/12/2020	2,10221
220P/McNaught	16/12/2020	1,55397
191P/McNaught	20/12/2020	2,04791
P/2010 T1 McNaught	21/12/2020	3,21265
98P/Takamizawa	10/01/2021	1,67346
277P/LINEAR	17/01/2021	1,91907
11P/Tempel-Swift-LINEAR	12/02/2021	1,58407
17P/Holmes	15/02/2021	2,05635
P/2007 B1 (Christensen)	15/02/2021	2,44277
P/2013 A2 Scotti	15/02/2021	2,17925
156P/Russell-LINEAR	16/02/2021	1,5931
193P/LINEAR-NEAT	17/02/2021	2,04444
246P/NEAT	17/02/2021	2,8678
28P/Neujmin 1	26/02/2021	1,55216
75D/Kohoutek	04/03/2021	1,78466
73P/Schwassmann-Wachmann 3-AZ	14/03/2021	0,94032
73P/Schwassmann-Wachmann 3-AY	15/03/2021	0,94019
265P/LINEAR	22/03/2021	1,50545
10P/Tempel 2	28/03/2021	1,42103
5D/Brorsen	28/03/2021	0,58985
73P/Schwassmann-Wachmann 3-AF	21/04/2021	0,94118
241P/LINEAR	01/05/2021	1,86108
73P/Schwassmann-Wachmann 3-AW	09/05/2021	0,94026
C/2011 W3 Lovejoy	14/05/2021	23,66474
P/2010 R2 La Sagra	14/05/2021	2,6192
7P/Pons-Winnecke	16/06/2021	1,2533
15P/Finlay	21/06/2021	0,97001
142P/Ge-Wang	02/07/2021	2,48792
C/2006 S5 (Hill)	13/07/2021	28,06305
120P/Mueller 1	14/07/2021	2,72904
P/2010 V1 Ikeya-Murakami	15/07/2021	1,5765
201P/LONEOS	16/07/2021	1,35801
252P/LINEAR	26/07/2021	1,00137
111P/Helin-Roman-Crockett	27/07/2021	3,7043
73P/Schwassmann-Wachmann 3-AA	27/07/2021	0,93928
8P/Tuttle	05/09/2021	1,02712
218P/LINEAR	06/09/2021	1,70272
52P/Harrington-Abell	09/09/2021	1,75711
6P/d'Arrest	09/09/2021	1,3535
73P/Schwassmann-Wachmann 3-BE	11/09/2021	0,93942
73P/Schwassmann-Wachmann 3-BH	16/09/2021	0,93982
73P/Schwassmann-Wachmann 3-BI	16/09/2021	0,93977
73P/Schwassmann-Wachmann 3-BM	16/09/2021	0,93976
73P/Schwassmann-Wachmann 3-BP	16/09/2021	0,93978
73P/Schwassmann-Wachmann 3-BA	18/09/2021	0,94025
P/2007 H1 (McNaught)	29/09/2021	2,28231
73P/Schwassmann-Wachmann 3-BB	07/10/2021	0,94242
57P/duToit-Neujmin-Delporte	16/10/2021	1,72471
P/2005 L1 (McNaught)	20/10/2021	3,14402
73P/Schwassmann-Wachmann 3-V	02/11/2021	0,91221
110P/Hartley 3	03/11/2021	2,48439
73P/Schwassmann-Wachmann 3-BD	09/11/2021	0,94616

Cometa-comet	Date	A.U.
106P/Schuster	12/11/2021	1,55612
73P/Schwassmann-Wachmann 3-BG	12/11/2021	0,93934
70P/Kojima	14/11/2021	2,01189
P/2008 QP20 (LINEAR-Hill)	16/11/2021	1,72332
P/2007 R4 Garradd	18/11/2021	1,92141
73P/Schwassmann-Wachmann 3-BR	23/11/2021	0,9396
P/2012 TK8 Tenagra	28/11/2021	3,0897
73P/Schwassmann-Wachmann 3-BF	08/12/2021	0,93957
172P/Yeung	13/12/2021	2,24099
181P/Shoemaker-Levy 6	17/12/2021	1,12749
4P/Faye	18/12/2021	1,66684
P/2005 JD108 (Catalina-NEAT)	28/12/2021	4,02909
P/2006 S4 (Christensen)	02/01/2022	3,06824
173P/Mueller 5	05/01/2022	4,21359
205P/Giacobini	05/01/2022	1,52638
73P/Schwassmann-Wachmann 3-AT	10/01/2022	0,93963
224P/LINEAR-NEAT	15/01/2022	1,88186
P/2009 U6 LINEAR	16/01/2022	1,4842
152P/Helin-Lawrence	22/01/2022	3,11624
67P/Churyumov-Gerasimenko	22/01/2022	1,24288
108P/Ciffreo	24/01/2022	1,71912
259P/Garradd	24/01/2022	1,79307
P/2008 R1 (Garradd)	26/01/2022	1,79306
9P/Tempel 1	27/01/2022	1,50908
221P/LINEAR	29/01/2022	1,79027
83D/Russell 1	06/02/2022	1,61154
P/2010 N1 WISE	07/02/2022	1,49452
P/2011 W1 PANSTARRS	07/02/2022	3,31093
230P/LINEAR	19/02/2022	1,486
97P/Metcalf-Brewington	20/02/2022	2,59765
182P/LONEOS	25/02/2022	0,97802
86P/Wild 3	18/03/2022	2,30113
73P/Schwassmann-Wachmann 3-U	20/03/2022	0,93939
179P/Jedicke	25/03/2022	4,08164
P/2009 L2 (Yang-Gao)	26/03/2022	1,29714
22P/Kopff	30/03/2022	1,57258
45P/Honda-Mrkos-Pajdusakova	31/03/2022	0,52976
19P/Borrelly	03/04/2022	1,35365
P/2013 J2 McNaught	06/04/2022	2,08685
274P/Tombaugh-Tenagra	09/04/2022	2,44178
P/2010 TO20 LINEAR-Grauer	11/04/2022	5,20178
P/2008 S1 (Catalina-McNaught)	12/04/2022	1,19052
73P/Schwassmann-Wachmann 3-AM	21/04/2022	0,93094
44P/Reinmuth 2	28/04/2022	2,1162
99P/Kowal 1	01/05/2022	4,74344
P/2012 O3 McNaught	02/05/2022	1,59931
73P/Schwassmann-Wachmann 3-BJ	05/05/2022	0,93925
73P/Schwassmann-Wachmann 3-AB	08/05/2022	0,93917
73P/Schwassmann-Wachmann 3-J	15/05/2022	0,93902
73P/Schwassmann-Wachmann 3-BK	18/05/2022	0,93927
73P/Schwassmann-Wachmann 3-BN	23/05/2022	0,93918
73P/Schwassmann-Wachmann 3-BO	25/05/2022	0,93922
113P/Spitaler	26/05/2022	2,12821
P/2007 R1 Larson	30/05/2022	4,35563

Cometa-comet	Date	A.U.
73P/Schwassmann-Wachmann 3-AX	02/06/2022	0,93801
135P/Shoemaker-Levy 8	06/06/2022	2,7211
73P/Schwassmann-Wachmann 3-N	07/06/2022	0,93925
73P/Schwassmann-Wachmann 3-BQ	11/06/2022	0,93915
73P/Schwassmann-Wachmann 3	12/06/2022	0,93277
238P/Read	17/06/2022	2,36467
73P/Schwassmann-Wachmann 3-AU	23/06/2022	0,9393
73P/Schwassmann-Wachmann 3-AK	24/06/2022	0,93771
P/2012 S2 La Sagra	27/06/2022	1,37467
P/2006 T1 (Levy)	30/06/2022	0,98959
39P/Oterma	01/07/2022	5,47115
73P/Schwassmann-Wachmann 3-AV	04/07/2022	0,93902
73P/Schwassmann-Wachmann 3-B	04/07/2022	0,93911
116P/Wild 4	06/07/2022	2,17519
73P/Schwassmann-Wachmann 3-E	06/07/2022	0,93744
73P/Schwassmann-Wachmann 3-C	07/07/2022	0,94286
73P/Schwassmann-Wachmann 3-G	09/07/2022	0,93918
117P/Helin-Roman-Alu 1	10/07/2022	3,05576
169P/NEAT	10/07/2022	0,60653
73P/Schwassmann-Wachmann 3-K	12/07/2022	0,93908
73P/Schwassmann-Wachmann 3-H	13/07/2022	0,93917
148P/Anderson-LINEAR	14/07/2022	1,70267
189P/NEAT	14/07/2022	1,17756
P/2009 Q1 Hill	17/07/2022	2,7887
73P/Schwassmann-Wachmann 3-AD	22/07/2022	0,94071
100P/Hartley 1	23/07/2022	1,99088
73P/Schwassmann-Wachmann 3-M	23/07/2022	0,9391
18D/Perrine-Mrkos	25/07/2022	1,27225
73P/Schwassmann-Wachmann 3-R	28/07/2022	0,93913
272P/NEAT	02/08/2022	2,4396
73P/Schwassmann-Wachmann 3-AO	02/08/2022	0,94075
73P/Schwassmann-Wachmann 3-Z	03/08/2022	0,94204
127P/Holt-Olmstead	07/08/2022	2,19355
P/2013 G4 PANSTARRS	07/08/2022	2,60366
73P/Schwassmann-Wachmann 3-AH	12/08/2022	0,94299
73P/Schwassmann-Wachmann 3-AN	15/08/2022	0,939
73P/Schwassmann-Wachmann 3-L	15/08/2022	0,93909
73P/Schwassmann-Wachmann 3-X	16/08/2022	0,93908
P/2011 Y1 Levy	16/08/2022	1,00745
255P/Levy	17/08/2022	1,00758
P/2008 Y2 (Gibbs)	29/08/2022	1,63839
73P/Schwassmann-Wachmann 3-AC	03/09/2022	0,93905
73P/Schwassmann-Wachmann 3-P	09/09/2022	0,93916
132P/Helin-Roman-Alu 2	10/09/2022	1,92416
214P/LINEAR	10/09/2022	1,83984
41P/Tuttle-Giacobini-Kresak	14/09/2022	1,04778
73P/Schwassmann-Wachmann 3-AG	15/09/2022	0,93904
73P/Schwassmann-Wachmann 3-AR	15/09/2022	0,9392
73P/Schwassmann-Wachmann 3-AQ	17/09/2022	0,93901
51P/Harrington	21/09/2022	1,68794
P/2007 S1 Zhao	26/09/2022	2,49438
73P/Schwassmann-Wachmann 3-BL	28/09/2022	0,9391
157P/Tritton	03/10/2022	1,35798
43P/Wolf-Harrington	04/10/2022	1,35756

Cometa-comet	Date	A.U.
73P/Schwassmann-Wachmann 3-AE	12/10/2022	0,93905
225P/LINEAR	17/10/2022	1,19206
61P/Shajn-Schaldach	18/10/2022	2,10838
263P/Gibbs	24/10/2022	1,2513
211P/Hill	25/10/2022	2,36164
196P/Tichy	28/10/2022	2,1516
73P/Schwassmann-Wachmann 3-W	07/11/2022	0,93913
32P/Comas Sola	10/11/2022	1,83366
73P/Schwassmann-Wachmann 3-AI	19/11/2022	0,9389
104P/Kowal 2	20/11/2022	1,3961
118P/Shoemaker-Levy 4	28/11/2022	1,98165
176P/LINEAR	30/11/2022	2,57553
73P/Schwassmann-Wachmann 3-BS	03/12/2022	0,93895
197P/LINEAR	08/12/2022	1,06146
204P/LINEAR-NEAT	13/12/2022	1,93852
P/2011 R3 Novichonok-Gerke	13/12/2022	3,55792
73P/Schwassmann-Wachmann 3-Q	14/12/2022	0,93972
81P/Wild 2	17/12/2022	1,59618
80P/Peters-Hartley	22/12/2022	1,62392
158P/Kowal-LINEAR	31/12/2022	4,58674
P/2011 Q3 McNaught	03/01/2023	2,36744
71P/Clark	06/01/2023	1,56927
244P/Scotti	08/01/2023	3,93928
96P/Machholz 1	09/01/2023	0,45518
73P/Schwassmann-Wachmann 3-AL	11/01/2023	0,93877
129P/Shoemaker-Levy 3	25/01/2023	3,91341
P/2013 CE31 MOSS	26/01/2023	4,01468
281P/MOSS	28/01/2023	4,01837
73P/Schwassmann-Wachmann 3-BC	28/01/2023	0,93881
119P/Parker-Hartley	04/02/2023	3,02697
256P/LINEAR	13/02/2023	2,68038
73P/Schwassmann-Wachmann 3-AS	16/02/2023	0,93882
73P/Schwassmann-Wachmann 3-AP	26/02/2023	0,93875
77P/Longmore	04/03/2023	2,31065
P/2000 Y3 (Scotti)	11/03/2023	4,01863
279P/La Sagra	19/04/2023	2,1477
170P/Christensen	04/05/2023	2,92844
94P/Russell 4	30/05/2023	2,23565
P/2006 R2 (Christensen)	24/06/2023	3,03924
180P/NEAT	26/06/2023	2,46868
280P/Larsen	05/07/2023	2,62041
185P/Petriew	14/07/2023	0,93184
P/2012 T1 PANSTARRS	15/07/2023	2,41916
P/2009 K1 (Gibbs)	16/07/2023	1,32261
126P/IRAS	24/07/2023	1,71781
147P/Kushida-Muramatsu	02/08/2023	2,75623
121P/Shoemaker-Holt 2	05/09/2023	3,75305
202P/Scotti	03/10/2023	2,52548
79P/du Toit-Hartley	05/10/2023	1,1239
103P/Hartley 2	07/10/2023	1,05869
2P/Encke	11/10/2023	0,33625
72D/Denning-Fujikawa	12/10/2023	0,77973
P/2010 A2 LINEAR	16/10/2023	2,00593
199P/Shoemaker 4	22/10/2023	2,93643

Cometa-comet	Date	A.U.
26P/Grigg-Skjellerup	29/12/2023	1,08589
P/2007 V2 Hill	29/12/2023	2,77478
251P/LINEAR	08/01/2024	1,7115
P/2007 T2 Kowalski	10/01/2024	0,69593
P/2011 NO1	18/01/2024	1,24493
216P/LINEAR	22/01/2024	2,1526
73P/Schwassmann-Wachmann 3-T	06/02/2024	0,93932
219P/LINEAR	17/02/2024	2,36394
213P/Van Ness	22/02/2024	2,12238
222P/LINEAR	28/02/2024	0,78105
207P/NEAT	29/02/2024	0,944
167P/CINEOS	08/03/2024	1,78354
C/2010 KW7 WISE	10/03/2024	27,84025
125P/Spacewatch	12/03/2024	1,52546
194P/LINEAR	21/03/2024	1,70702
12P/Pons-Brooks	30/03/2024	1,18375
150P/LONEOS	31/03/2024	1,76286
227P/Catalina-LINEAR	07/04/2024	1,79592
144P/Kushida	11/04/2024	1,43901
P/2009 S1 Gibbs	16/04/2024	2,41458
226P/Pigott-LINEAR-Kowalski	06/05/2024	1,91685
237P/LINEAR	14/05/2024	2,41934
50P/Arend	14/05/2024	1,92373
209P/LINEAR	23/05/2024	0,91244
P/2011 VJ5 Lemmon	26/05/2024	1,50578
46P/Wirtanen	28/05/2024	1,05741
89P/Russell 2	30/05/2024	2,27994
P/2010 T2 PANSTARRS	10/06/2024	3,75285
154P/Brewington	15/06/2024	1,59037
212P/NEAT	27/06/2024	1,65423
P/2010 WK LINEAR	17/07/2024	1,76625
146P/Shoemaker-LINEAR	19/07/2024	1,41774
P/1998 QP54 (LONEOS-Tucker)	20/07/2024	1,87992
267P/LONEOS	27/07/2024	1,33776
208P/McMillan	31/07/2024	2,52521
P/2012 T2 PANSTARRS	13/08/2024	4,88473
54P/de Vico-Swift-NEAT	01/09/2024	2,17176
37P/Forbes	04/09/2024	1,57822
P/2004 DO29 (Spacewatch-LINEAR)	11/09/2024	4,09276
73P/Schwassmann-Wachmann 3-S	12/09/2024	1,02237
268P/Bernardi	16/09/2024	2,34535
33P/Daniel	29/09/2024	2,16951
65P/Gunn	09/10/2024	2,62535
25D/Neujmin 2	10/10/2024	1,33817
73P/Schwassmann-Wachmann 3-Y	24/10/2024	0,93917
253P/PANSTARRS	26/10/2024	2,03802
62P/Tsuchinshan 1	27/10/2024	1,48905
130P/McNaught-Hughes	29/10/2024	2,08467
P/2011 A2 Scotti	04/11/2024	1,58639
P/2005 RV25 (LONEOS-Christensen)	07/11/2024	3,60784
P/2012 US27 Siding Spring	18/11/2024	1,82065
C/2010 G2 Hill	23/11/2024	28,02465
234P/LINEAR	24/11/2024	2,85633
P/2011 R2 PANSTARRS	06/12/2024	2,04923

Cometa-comet	Date	A.U.
30P/Reinmuth 1	20/12/2024	1,88319
190P/Mueller	21/12/2024	2,03575
136P/Mueller 3	26/12/2024	2,96302
13P/Olbers	04/01/2025	0,93857
105P/Singer Brewster	30/01/2025	2,05052
236P/LINEAR	30/01/2025	1,83108
P/2011 U2 Bressi	30/01/2025	4,83743
P/2011 UA134 Spacewatch-PANSTARR	11/02/2025	2,05143
229P/Gibbs	21/02/2025	2,44021
P/2010 A3 Hill	26/02/2025	1,62179
249P/LINEAR	03/03/2025	0,51059
P/2010 K2 WISE	11/03/2025	1,19683
192P/Shoemaker-Levy 1	05/04/2025	1,53239
49P/Arend-Rigaux	05/04/2025	1,42434
P/2007 R3 Gibbs	06/04/2025	2,50361
P/2010 H2 Vales	07/04/2025	3,10566
250P/Larson	12/04/2025	2,21403
21P/Giacobini-Zinner	19/04/2025	1,03051
217P/LINEAR	05/05/2025	1,22348
164P/Christensen	11/05/2025	1,67361
276P/Vorobjov	25/06/2025	3,92276
60P/Tsuchinshan 2	29/06/2025	1,61816
P/2010 B2 WISE	04/07/2025	1,60697
P/2010 P4 WISE	19/07/2025	1,86247
P/2008 T1 (Boattini)	25/07/2025	3,04484
P/2008 A2 LINEAR	28/07/2025	1,30541
3D/Biela	17/08/2025	0,87907
198P/ODAS	06/09/2025	1,9782
48P/Johnson	12/09/2025	2,29998
40P/Vaisala 1	14/09/2025	1,79597
171P/Spahr	20/09/2025	1,76498
47P/Ashbrook-Jackson	18/10/2025	2,80981
P/2012 F5 Gibbs	28/10/2025	2,87264
42P/Neujmin 3	11/12/2025	2,01471
240P/NEAT	19/12/2025	2,12638
P/2012 O1 McNaught	21/12/2025	1,4989
P/1999 XN120 (Catalina)	24/12/2025	3,30298
141P/Machholz 2-D	13/01/2026	0,74901
233P/La Sagra	27/01/2026	1,79498
24P/Schaumasse	28/01/2026	1,20501
141P/Machholz 2-A	29/01/2026	0,75285
210P/Christensen	05/02/2026	0,54642
73P/Schwassmann-Wachmann 3-AZ	14/02/2026	0,94032
P/2012 O2 McNaught	14/02/2026	1,66083
73P/Schwassmann-Wachmann 3-AY	15/02/2026	0,94019
131P/Mueller 2	20/02/2026	2,41675
162P/Siding Spring	23/02/2026	1,22767
245P/WISE	01/03/2026	2,14099
243P/NEAT	08/03/2026	2,45754
188P/LINEAR-Mueller	19/03/2026	2,55256
235P/LINEAR	21/03/2026	2,74091
12P/Pons-Brooks	22/03/2026	1,191
138P/Shoemaker-Levy 7	25/03/2026	1,70034
88P/Howell	25/03/2026	1,36157

Cometa-comet	Date	A.U.
P/2009 Q4 Boattini	30/03/2026	1,31873
206P/Barnard-Boattini	04/04/2026	1,13648
73P/Schwassmann-Wachmann 3-AF	06/04/2026	0,94118
P/2011 V1 Boattini	10/04/2026	1,71167
76P/West-Kohoutek-Ikemura	13/04/2026	1,60024
P/2008 Q2 (Ory)	19/04/2026	1,38223
73P/Schwassmann-Wachmann 3-AW	30/04/2026	0,94026
261P/Larson	01/05/2026	2,18969
124P/Mrkos	09/05/2026	1,64488
93P/Lovas 1	13/05/2026	1,7041
P/2010 D1 WISE	19/05/2026	2,66908
34D/Gale	25/05/2026	1,18291
128P/Shoemaker-Holt 1-A	31/05/2026	3,04728
P/2009 S2 McNaught	16/06/2026	2,2036
220P/McNaught	17/06/2026	1,55397
P/2010 J5 McNaught	20/06/2026	3,74762
78P/Gehrels 2	21/06/2026	2,0085
P/2004 FY140 (LINEAR)	10/07/2026	4,1076
168P/Hergenrother	20/07/2026	1,41484
259P/Garradd	26/07/2026	1,79307
74P/Smirnova-Chernykh	26/07/2026	3,54877
215P/NEAT	27/07/2026	3,21596
175P/Hergenrother	28/07/2026	2,05472
P/2008 R1 (Garradd)	29/07/2026	1,79306
10P/Tempel 2	08/08/2026	1,42103
73P/Schwassmann-Wachmann 3-AA	12/08/2026	0,93928
145P/Shoemaker-Levy 5	14/08/2026	1,987
128P/Shoemaker-Holt 1-B	15/08/2026	3,06883
14P/Wolf	22/08/2026	2,7248
P/2011 W2 Rinner	27/08/2026	2,30311
123P/West-Hartley	05/09/2026	2,12871
5D/Brorsen	13/09/2026	0,58985
114P/Wiseman-Skiff	15/09/2026	1,57484
16P/Brooks 2	17/09/2026	1,46657
73P/Schwassmann-Wachmann 3-AJ	18/09/2026	0,93608
169P/NEAT	22/09/2026	0,60653
P/2010 L1 WISE	24/09/2026	2,1489
P/2007 C1 (Christensen)	27/09/2026	2,05079
73P/Schwassmann-Wachmann 3-BE	13/10/2026	0,93942
73P/Schwassmann-Wachmann 3-BM	19/10/2026	0,93976
73P/Schwassmann-Wachmann 3-BP	19/10/2026	0,93978
P/2007 R2 Gibbs	19/10/2026	1,46589
73P/Schwassmann-Wachmann 3-BH	20/10/2026	0,93982
73P/Schwassmann-Wachmann 3-BI	20/10/2026	0,93977
73P/Schwassmann-Wachmann 3-BA	22/10/2026	0,94025
P/2010 R2 La Sagra	23/10/2026	2,6192
112P/Urata-Niijima	29/10/2026	1,46469
69P/Taylor	02/11/2026	2,27285
260P/McNaught	06/11/2026	1,4979
163P/NEAT	16/11/2026	2,05709
73P/Schwassmann-Wachmann 3-BB	18/11/2026	0,94242
82P/Gehrels 3	22/11/2026	3,63331
252P/LINEAR	29/11/2026	1,00137
P/2010 V1 Ikeya-Murakami	30/11/2026	1,5765

Cometa-comet	Date	A.U.
266P/Christensen	08/12/2026	2,32578
P/2006 U5 (Christensen)	11/12/2026	2,3259
13P/Olbers	14/12/2026	0,94414
247P/LINEAR	17/12/2026	1,49615
73P/Schwassmann-Wachmann 3-V	21/12/2026	0,91221
73P/Schwassmann-Wachmann 3-BD	01/01/2027	0,94616
73P/Schwassmann-Wachmann 3-BG	04/01/2027	0,93934
87P/Bus	12/01/2027	2,17342
73P/Schwassmann-Wachmann 3-BR	19/01/2027	0,9396
2P/Encke	26/01/2027	0,33625
P/2012 K3 Gibbs	26/01/2027	2,03149
73P/Schwassmann-Wachmann 3-BF	07/02/2027	0,93957
182P/LONEOS	04/03/2027	0,97802
P/2008 J2 Beshore	12/03/2027	2,45485
73P/Schwassmann-Wachmann 3-AT	25/03/2027	0,93963
264P/Larsen	26/03/2027	2,43729
184P/Lovas 2	31/03/2027	1,45201
P/2010 A2 LINEAR	05/04/2027	2,00593
P/2007 H3 (Garradd)	07/04/2027	1,82927
143P/Kowal-Mrkos	19/04/2027	2,53815
P/2004 VR8 (LONEOS)	11/05/2027	2,37653
223P/Skiff	19/06/2027	2,41059
73P/Schwassmann-Wachmann 3-U	23/06/2027	0,93939
11P/Tempel-Swift-LINEAR	28/06/2027	1,58407
45P/Honda-Mrkos-Pajdusakova	02/07/2027	0,52976
84P/Giclas	03/07/2027	1,85166
139P/Vaisala-Oterma	06/07/2027	3,40275
183P/Korlevic-Juric	06/07/2027	3,89437
189P/NEAT	11/07/2027	1,17756
231P/LINEAR-NEAT	11/07/2027	3,03121
187P/LINEAR	22/07/2027	3,69532
P/1998 VS24 (LINEAR)	28/07/2027	3,42241
92P/Sanguin	04/08/2027	1,80804
73P/Schwassmann-Wachmann 3-AM	05/08/2027	0,93094
9P/Tempel 1	06/08/2027	1,50908
191P/McNaught	10/08/2027	2,04791
73P/Schwassmann-Wachmann 3-BJ	23/08/2027	0,93925
P/2008 T4 (Hill)	26/08/2027	2,51086
73P/Schwassmann-Wachmann 3-AB	28/08/2027	0,93917
73P/Schwassmann-Wachmann 3-J	06/09/2027	0,93902
178P/Hug-Bell	10/09/2027	1,94694
73P/Schwassmann-Wachmann 3-BK	10/09/2027	0,93927
193P/LINEAR-NEAT	11/09/2027	2,04444
73P/Schwassmann-Wachmann 3-BN	17/09/2027	0,93918
73P/Schwassmann-Wachmann 3-BO	19/09/2027	0,93922
P/2007 T6 Catalina	23/09/2027	2,23777
P/2006 T1 (Levy)	27/09/2027	0,9896
73P/Schwassmann-Wachmann 3-AX	30/09/2027	0,93801
P/2006 G1 (McNaught)	01/10/2027	2,63209
278P/McNaught	05/10/2027	2,0867
73P/Schwassmann-Wachmann 3-N	07/10/2027	0,93925
73P/Schwassmann-Wachmann 3-BQ	12/10/2027	0,93915
218P/LINEAR	15/10/2027	1,70272
73P/Schwassmann-Wachmann 3	15/10/2027	0,93277

Cometa-comet	Date	A.U.
197P/LINEAR	17/10/2027	1,06146
P/2010 A1 Hill	22/10/2027	1,95063
7P/Pons-Winnecke	25/10/2027	1,2533
73P/Schwassmann-Wachmann 3-AU	28/10/2027	0,9393
73P/Schwassmann-Wachmann 3-AK	29/10/2027	0,9377
75D/Kohoutek	04/11/2027	1,78466
P/2010 N1 WISE	05/11/2027	1,49452
P/2006 XG16 (Spacewatch)	11/11/2027	2,10221
73P/Schwassmann-Wachmann 3-AV	12/11/2027	0,93902
73P/Schwassmann-Wachmann 3-B	12/11/2027	0,93911
73P/Schwassmann-Wachmann 3-E	15/11/2027	0,93744
73P/Schwassmann-Wachmann 3-C	17/11/2027	0,94286
73P/Schwassmann-Wachmann 3-G	18/11/2027	0,93918
73P/Schwassmann-Wachmann 3-K	22/11/2027	0,93908
73P/Schwassmann-Wachmann 3-H	24/11/2027	0,93917
P/2011 Y1 Levy	01/12/2027	1,00745
255P/Levy	02/12/2027	1,00758
73P/Schwassmann-Wachmann 3-AD	06/12/2027	0,94071
73P/Schwassmann-Wachmann 3-M	07/12/2027	0,9391
P/2009 W1 Hill	12/12/2027	2,93817
73P/Schwassmann-Wachmann 3-R	13/12/2027	0,93913
156P/Russell-LINEAR	18/12/2027	1,5931
257P/Catalina	18/12/2027	2,12595
P/2005 JY126 (Catalina)	18/12/2027	2,12588
15P/Finlay	21/12/2027	0,97001
73P/Schwassmann-Wachmann 3-AO	21/12/2027	0,94075
73P/Schwassmann-Wachmann 3-Z	22/12/2027	0,94204
167P/CINEOS	02/01/2028	1,78354
73P/Schwassmann-Wachmann 3-AH	03/01/2028	0,94299
201P/LONEOS	05/01/2028	1,35801
17P/Holmes	07/01/2028	2,05635
73P/Schwassmann-Wachmann 3-AN	07/01/2028	0,939
73P/Schwassmann-Wachmann 3-L	07/01/2028	0,93909
73P/Schwassmann-Wachmann 3-X	08/01/2028	0,93908
73P/Schwassmann-Wachmann 3-AC	31/01/2028	0,93905
263P/Gibbs	01/02/2028	1,2513
238P/Read	02/02/2028	2,36467
73P/Schwassmann-Wachmann 3-P	09/02/2028	0,93916
41P/Tuttle-Giacobini-Kresak	15/02/2028	1,04778
102P/Shoemaker 1	16/02/2028	1,97384
56P/Slaughter-Burnham	16/02/2028	2,53497
73P/Schwassmann-Wachmann 3-AG	17/02/2028	0,93904
73P/Schwassmann-Wachmann 3-AR	17/02/2028	0,9392
73P/Schwassmann-Wachmann 3-AQ	20/02/2028	0,93901
224P/LINEAR-NEAT	25/02/2028	1,88186
64P/Swift-Gehrels	26/02/2028	1,37711
73P/Schwassmann-Wachmann 3-BL	06/03/2028	0,9391
149P/Mueller 4	07/03/2028	2,6509
57P/duToit-Neujmin-Delporte	11/03/2028	1,72471
83D/Russell 1	13/03/2028	1,61154
59P/Kearns-Kwee	17/03/2028	2,35521
6P/d'Arrest	23/03/2028	1,3535
73P/Schwassmann-Wachmann 3-AE	24/03/2028	0,93905
31P/Schwassmann-Wachmann 2	31/03/2028	3,42307

Cometa-comet	Date	A.U.
P/2009 U6 LINEAR	05/04/2028	1,4842
P/2012 SB6 Lemmon	17/04/2028	2,40608
96P/Machholz 1	23/04/2028	0,65504
73P/Schwassmann-Wachmann 3-W	28/04/2028	0,93912
73P/Schwassmann-Wachmann 3-AI	13/05/2028	0,9389
P/2008 QP20 (LINEAR-Hill)	25/05/2028	1,72332
230P/LINEAR	27/05/2028	1,486
73P/Schwassmann-Wachmann 3-BS	02/06/2028	0,93895
239P/LINEAR	05/06/2028	1,64506
P/2008 O2 (McNaught)	07/06/2028	3,79833
73P/Schwassmann-Wachmann 3-Q	16/06/2028	0,93972
137P/Shoemaker-Levy 2	17/06/2028	1,90606
98P/Takamizawa	17/06/2028	1,67346
P/2006 F1 (Kowalski)	28/06/2028	4,11995
67P/Churyumov-Gerasimenko	04/07/2028	1,24288
P/2010 U2 Hill	11/07/2028	2,55374
172P/Yeung	14/07/2028	2,24099
71P/Clark	18/07/2028	1,56927
160P/LINEAR	22/07/2028	2,07021
73P/Schwassmann-Wachmann 3-AL	23/07/2028	0,93877
91P/Russell 3	24/07/2028	2,61655
221P/LINEAR	31/07/2028	1,79027
176P/LINEAR	16/08/2028	2,57553
73P/Schwassmann-Wachmann 3-BC	16/08/2028	0,93881
228P/LINEAR	19/08/2028	3,42808
277P/LINEAR	24/08/2028	1,91907
P/2009 L2 (Yang-Gao)	26/08/2028	1,29714
22P/Kopff	31/08/2028	1,57258
205P/Giacobini	02/09/2028	1,52638
73P/Schwassmann-Wachmann 3-AS	10/09/2028	0,93882
110P/Hartley 3	18/09/2028	2,48439
58P/Jackson-Neujmin	24/09/2028	1,38117
73P/Schwassmann-Wachmann 3-AP	24/09/2028	0,93875
232P/Hill	25/09/2028	2,98319
P/2011 U1 PANSTARRS	01/10/2028	2,35619
P/2007 H1 (McNaught)	20/10/2028	2,28231
79P/du Toit-Hartley	25/10/2028	1,1239
P/2012 T1 PANSTARRS	08/11/2028	2,41916
100P/Hartley 1	14/11/2028	1,99088
43P/Wolf-Harrington	20/11/2028	1,35756
70P/Kojima	04/12/2028	2,01189
185P/Petriew	27/12/2028	0,93184
222P/LINEAR	27/12/2028	0,78105
116P/Wild 4	29/12/2028	2,17519
127P/Holt-Olmstead	29/12/2028	2,19355
36P/Whipple	15/01/2029	3,08188
P/2008 S1 (Catalina-McNaught)	16/01/2029	1,19052
157P/Tritton	22/01/2029	1,35798
104P/Kowal 2	23/01/2029	1,3961
19P/Borrelly	07/02/2029	1,35365
86P/Wild 3	15/02/2029	2,30113
P/2013 A2 Scotti	23/02/2029	2,17925
106P/Schuster	04/03/2029	1,55612
246P/NEAT	11/03/2029	2,8678

Cometa-comet	Date	A.U.
52P/Harrington-Abell	24/03/2029	1,75711
26P/Grigg-Skjellerup	27/03/2029	1,08589
18D/Perrine-Mrkos	12/04/2029	1,27225
108P/Ciffreo	29/04/2029	1,71912
115P/Maury	07/05/2029	2,0397
118P/Shoemaker-Levy 4	11/05/2029	1,98165
81P/Wild 2	15/05/2029	1,59618
44P/Reinmuth 2	02/06/2029	2,1162
209P/LINEAR	05/06/2029	0,91244
186P/Garradd	15/06/2029	4,26391
P/2008 Y2 (Gibbs)	17/06/2029	1,63839
P/2007 T2 Kowalski	18/06/2029	0,69593
P/2012 G1 PANSTARRS	19/06/2029	2,58417
225P/LINEAR	21/06/2029	1,19206
113P/Spitaler	27/06/2029	2,12821
181P/Shoemaker-Levy 6	29/06/2029	1,12749
4P/Faye	05/07/2029	1,66684
214P/LINEAR	14/07/2029	1,83984
211P/Hill	20/07/2029	2,36164
148P/Anderson-LINEAR	08/08/2029	1,70267
258P/PANSTARRS	27/08/2029	3,4792
P/2012 H1 PANSTARRS	07/09/2029	3,47997
125P/Spacewatch	23/09/2029	1,52546
P/2005 L1 (McNaught)	23/09/2029	3,14402
249P/LINEAR	18/10/2029	0,51059
46P/Wirtanen	05/11/2029	1,05741
51P/Harrington	07/11/2029	1,68794
61P/Shajn-Schaldach	07/11/2029	2,10838
P/2008 Y1 (Boattini)	11/11/2029	1,27061
120P/Mueller 1	03/12/2029	2,72904
135P/Shoemaker-Levy 8	04/12/2029	2,7211
204P/LINEAR-NEAT	14/12/2029	1,93852
73P/Schwassmann-Wachmann 3-T	27/12/2029	0,93932
265P/LINEAR	28/12/2029	1,50545
94P/Russell 4	29/12/2029	2,23565
77P/Longmore	30/12/2029	2,31065
111P/Helin-Roman-Crockett	20/01/2030	3,7043
279P/La Sagra	22/01/2030	2,1477
P/2007 V1 Larson	31/01/2030	2,67658
P/2010 K2 WISE	31/01/2030	1,19683
P/2010 A5 LINEAR	11/02/2030	1,69746
203P/Korlevic	12/02/2030	3,1824
P/2007 S1 Zhao	20/02/2030	2,49438
196P/Tichy	07/03/2030	2,15161
25D/Neujmin 2	16/03/2030	1,33817
103P/Hartley 2	28/03/2030	1,05869
P/2009 SK280 Spacewatch-Hill	03/04/2030	4,20758
2P/Encke	14/05/2030	0,33625
P/2012 TK8 Tenagra	19/06/2030	3,0897
200P/Larsen	23/06/2030	3,28038
213P/Van Ness	27/06/2030	2,12238
P/2007 T4 Gibbs	01/07/2030	1,99968
267P/LONEOS	16/07/2030	1,33776
251P/LINEAR	17/07/2030	1,7115

```
P/2009 K1 (Gibbs)                    25/07/2030    1,32261
P/2011 VJ5 Lemmon                    20/08/2030    1,50578
P/2010 B2 WISE                       08/09/2030    1,60697
68P/Klemola                          20/09/2030    1,75902
P/2010 A2 LINEAR                     22/09/2030    2,00593
73P/Schwassmann-Wachmann 3-S         15/10/2030    1,02237
117P/Helin-Roman-Alu 1               22/10/2030    3,05576
P/2013 J2 McNaught                   14/11/2030    2,08685
169P/NEAT                            05/12/2030    0,60653
73P/Schwassmann-Wachmann 3-Y         10/12/2030    0,93917
132P/Helin-Roman-Alu 2               23/12/2030    1,92416
```

MAGNITUDINE MINIMA<12
MINIMUM MAGNITUDE<12

Comet	Date	Mag
C/2011 F1 LINEAR	20/01/2013	9,8
C/2012 T5 Bressi	24/02/2013	7,9
C/2011 L4 PANSTARRS	10/03/2013	0,5
C/2012 F6 Lemmon	18/03/2013	9,4
112P/Urata-Niijima	04/07/2013	10,1
46P/Wirtanen	11/07/2013	10,9
C/2013 G5 Catalina	11/09/2013	11,8
154P/Brewington	17/10/2013	7,7
2P/Encke	20/11/2013	4,6
C/2012 S1 ISON	28/11/2013	-5,6
25D/Neujmin 2	05/12/2013	9,3
32P/Comas Sola	04/01/2014	11,5
C/2012 X1 LINEAR	24/02/2014	11,6
124P/Mrkos	26/03/2014	8,6
210P/Christensen	31/08/2014	11,2
C/2013 A1 Siding Spring	10/09/2014	7,7
72D/Denning-Fujikawa	11/10/2014	-0,9
C/2012 K1 PANSTARRS	14/11/2014	6
34D/Gale	26/05/2015	5,7
19P/Borrelly	28/05/2015	9,7
141P/Machholz 2-D	03/08/2015	8,6
141P/Machholz 2-A	11/08/2015	2
5D/Brorsen	09/10/2015	-0,3
22P/Kopff	18/10/2015	9,7
10P/Tempel 2	08/11/2015	10,1
18D/Perrine-Mrkos	11/11/2015	5
P/2010 V1 Ikeya-Murakami	27/02/2016	8,9
73P/Schwassmann-Wachmann 3-AZ	04/04/2016	11,2
73P/Schwassmann-Wachmann 3-AY	06/04/2016	11,2
252P/LINEAR	07/04/2016	-0,7
73P/Schwassmann-Wachmann 3-AF	24/04/2016	10,2
73P/Schwassmann-Wachmann 3-AW	01/05/2016	9,3
118P/Shoemaker-Levy 4	09/06/2016	10,8
167P/CINEOS	26/06/2016	10,3
9P/Tempel 1	01/07/2016	10,2
81P/Wild 2	08/07/2016	11,8
73P/Schwassmann-Wachmann 3-AA	29/07/2016	8,5
73P/Schwassmann-Wachmann 3-AJ	10/08/2016	9,9
73P/Schwassmann-Wachmann 3-BE	19/08/2016	10,6
43P/Wolf-Harrington	22/08/2016	11,5
73P/Schwassmann-Wachmann 3-BH	22/08/2016	10,7
73P/Schwassmann-Wachmann 3-BI	22/08/2016	10,7
73P/Schwassmann-Wachmann 3-BM	22/08/2016	10,7
73P/Schwassmann-Wachmann 3-BP	22/08/2016	10,7
73P/Schwassmann-Wachmann 3-BA	23/08/2016	10,8
73P/Schwassmann-Wachmann 3-BB	03/09/2016	11,2
73P/Schwassmann-Wachmann 3-V	19/09/2016	11,4
73P/Schwassmann-Wachmann 3-BD	22/09/2016	11,8
73P/Schwassmann-Wachmann 3-BG	25/09/2016	11,8
73P/Schwassmann-Wachmann 3-BR	01/10/2016	11,9
45P/Honda-Mrkos-Pajdusakova	04/01/2017	7
73P/Schwassmann-Wachmann 3-M	05/03/2017	11,9
73P/Schwassmann-Wachmann 3-R	07/03/2017	11,9
2P/Encke	09/03/2017	3,5

Comet	Date	Mag
73P/Schwassmann-Wachmann 3-AO	11/03/2017	11,8
73P/Schwassmann-Wachmann 3-Z	11/03/2017	11,8
73P/Schwassmann-Wachmann 3-AH	17/03/2017	11,7
73P/Schwassmann-Wachmann 3-AN	19/03/2017	11,7
73P/Schwassmann-Wachmann 3-L	19/03/2017	11,7
73P/Schwassmann-Wachmann 3-X	19/03/2017	11,6
73P/Schwassmann-Wachmann 3-AC	30/03/2017	11,4
73P/Schwassmann-Wachmann 3-P	03/04/2017	11,3
73P/Schwassmann-Wachmann 3-AG	06/04/2017	11,1
73P/Schwassmann-Wachmann 3-AR	06/04/2017	11,2
P/2006 T1 (Levy)	06/04/2017	11,3
73P/Schwassmann-Wachmann 3-AQ	07/04/2017	11,1
41P/Tuttle-Giacobini-Kresak	10/04/2017	6,7
73P/Schwassmann-Wachmann 3-BL	14/04/2017	10,8
103P/Hartley 2	18/04/2017	11,5
73P/Schwassmann-Wachmann 3-AE	21/04/2017	10,5
73P/Schwassmann-Wachmann 3-W	01/05/2017	9,4
P/2011 Y1 Levy	03/05/2017	11,8
73P/Schwassmann-Wachmann 3-AI	04/05/2017	8,5
73P/Schwassmann-Wachmann 3-BS	10/05/2017	6,9
73P/Schwassmann-Wachmann 3-Q	24/05/2017	4,5
71P/Clark	16/06/2017	11,5
263P/Gibbs	15/07/2017	9,2
73P/Schwassmann-Wachmann 3-AL	15/07/2017	7,2
73P/Schwassmann-Wachmann 3-BC	30/07/2017	8,6
73P/Schwassmann-Wachmann 3-AS	07/08/2017	9,7
73P/Schwassmann-Wachmann 3-AP	12/08/2017	10,1
96P/Machholz 1	28/09/2017	10,6
24P/Schaumasse	30/10/2017	10,5
73P/Schwassmann-Wachmann 3-T	14/03/2018	11,8
66P/du Toit	11/05/2018	7,6
73P/Schwassmann-Wachmann 3-S	18/08/2018	11,5
38P/Stephan-Oterma	01/09/2018	11,5
73P/Schwassmann-Wachmann 3-Y	14/09/2018	11,5
21P/Giacobini-Zinner	17/09/2018	6,6
64P/Swift-Gehrels	25/10/2018	11
3D/Biela	27/11/2018	0
46P/Wirtanen	27/12/2018	5,5
239P/LINEAR	12/01/2019	9,8
25D/Neujmin 2	30/04/2019	8,7
209P/LINEAR	07/06/2019	11,3
155P/Shoemaker 3	13/01/2020	11,1
112P/Urata-Niijima	19/02/2020	9,8
124P/Mrkos	29/03/2020	9
210P/Christensen	23/05/2020	11,8
2P/Encke	26/06/2020	4,7
249P/LINEAR	13/07/2020	11,2
167P/CINEOS	26/07/2020	10,5
85P/Boethin	28/09/2020	6,1
141P/Machholz 2-D	24/10/2020	5,4
162P/Siding Spring	26/10/2020	10,1
141P/Machholz 2-A	15/11/2020	-1,1
73P/Schwassmann-Wachmann 3-AZ	10/03/2021	11,9
73P/Schwassmann-Wachmann 3-AY	11/03/2021	11,8

Comet	Date	Mag
5D/Brorsen	30/03/2021	-0,8
10P/Tempel 2	31/03/2021	10,5
73P/Schwassmann-Wachmann 3-AF	13/04/2021	10,9
73P/Schwassmann-Wachmann 3-AW	26/04/2021	10
7P/Pons-Winnecke	20/06/2021	8,4
P/2010 V1 Ikeya-Murakami	30/06/2021	11,9
252P/LINEAR	27/07/2021	6,3
73P/Schwassmann-Wachmann 3-AA	09/08/2021	9,8
6P/d'Arrest	02/09/2021	11,9
8P/Tuttle	07/09/2021	9,5
73P/Schwassmann-Wachmann 3-BE	16/09/2021	11,6
73P/Schwassmann-Wachmann 3-BH	21/09/2021	11,7
73P/Schwassmann-Wachmann 3-BI	21/09/2021	11,7
73P/Schwassmann-Wachmann 3-BM	21/09/2021	11,7
73P/Schwassmann-Wachmann 3-BP	21/09/2021	11,7
73P/Schwassmann-Wachmann 3-BA	22/09/2021	11,7
P/2008 QP20 (LINEAR-Hill)	23/10/2021	11,8
4P/Faye	04/11/2021	11,3
9P/Tempel 1	04/02/2022	11,6
83D/Russell 1	24/02/2022	11,6
73P/Schwassmann-Wachmann 3-U	16/03/2022	11,7
19P/Borrelly	31/03/2022	9,4
45P/Honda-Mrkos-Pajdusakova	01/04/2022	8,7
22P/Kopff	11/04/2022	9,5
73P/Schwassmann-Wachmann 3-AM	13/04/2022	10,8
73P/Schwassmann-Wachmann 3-BJ	24/04/2022	10,3
73P/Schwassmann-Wachmann 3-AB	26/04/2022	10,1
73P/Schwassmann-Wachmann 3-J	30/04/2022	9,7
73P/Schwassmann-Wachmann 3-BK	01/05/2022	9,5
73P/Schwassmann-Wachmann 3-BN	03/05/2022	9
73P/Schwassmann-Wachmann 3-BO	04/05/2022	8,8
73P/Schwassmann-Wachmann 3-AX	08/05/2022	7,3
73P/Schwassmann-Wachmann 3-N	15/05/2022	5,7
73P/Schwassmann-Wachmann 3-BQ	22/05/2022	4,5
73P/Schwassmann-Wachmann 3	23/05/2022	4,6
73P/Schwassmann-Wachmann 3-AU	25/06/2022	6,3
73P/Schwassmann-Wachmann 3-AK	27/06/2022	6,5
73P/Schwassmann-Wachmann 3-AV	21/07/2022	7,6
73P/Schwassmann-Wachmann 3-B	21/07/2022	7,6
73P/Schwassmann-Wachmann 3-E	25/07/2022	7,9
18D/Perrine-Mrkos	26/07/2022	8,9
73P/Schwassmann-Wachmann 3-C	26/07/2022	8,1
73P/Schwassmann-Wachmann 3-G	28/07/2022	8,2
73P/Schwassmann-Wachmann 3-K	30/07/2022	8,5
73P/Schwassmann-Wachmann 3-H	01/08/2022	8,7
73P/Schwassmann-Wachmann 3-AD	06/08/2022	9,5
73P/Schwassmann-Wachmann 3-M	07/08/2022	9,6
73P/Schwassmann-Wachmann 3-R	10/08/2022	9,9
73P/Schwassmann-Wachmann 3-AO	14/08/2022	10,2
73P/Schwassmann-Wachmann 3-Z	14/08/2022	10,3
P/2011 Y1 Levy	14/08/2022	11,9
73P/Schwassmann-Wachmann 3-AH	21/08/2022	10,7
73P/Schwassmann-Wachmann 3-AN	24/08/2022	10,8
73P/Schwassmann-Wachmann 3-L	24/08/2022	10,8

Comet	Date	Mag
73P/Schwassmann-Wachmann 3-X	24/08/2022	10,8
73P/Schwassmann-Wachmann 3-AC	09/09/2022	11,4
73P/Schwassmann-Wachmann 3-P	14/09/2022	11,5
73P/Schwassmann-Wachmann 3-AG	20/09/2022	11,6
73P/Schwassmann-Wachmann 3-AR	20/09/2022	11,6
73P/Schwassmann-Wachmann 3-AQ	22/09/2022	11,7
73P/Schwassmann-Wachmann 3-BL	02/10/2022	11,9
43P/Wolf-Harrington	15/10/2022	10,7
263P/Gibbs	26/10/2022	8,7
118P/Shoemaker-Levy 4	31/12/2022	8,6
81P/Wild 2	13/01/2023	11,4
103P/Hartley 2	04/10/2023	8,3
2P/Encke	10/10/2023	4,9
72D/Denning-Fujikawa	15/10/2023	-1
46P/Wirtanen	28/05/2024	10,9
154P/Brewington	19/06/2024	10,4
167P/CINEOS	25/08/2024	11,5
25D/Neujmin 2	12/10/2024	9,9
21P/Giacobini-Zinner	18/04/2025	10,7
3D/Biela	16/08/2025	4,9
141P/Machholz 2-D	16/01/2026	8,4
24P/Schaumasse	26/01/2026	6,9
141P/Machholz 2-A	01/02/2026	2,4
210P/Christensen	07/03/2026	10
73P/Schwassmann-Wachmann 3-AF	31/03/2026	11,4
124P/Mrkos	07/04/2026	9,5
73P/Schwassmann-Wachmann 3-AW	20/04/2026	10,5
34D/Gale	23/05/2026	5,9
10P/Tempel 2	03/08/2026	6,9
73P/Schwassmann-Wachmann 3-AA	21/08/2026	10,6
5D/Brorsen	12/09/2026	0
73P/Schwassmann-Wachmann 3-AJ	23/09/2026	11,7
16P/Brooks 2	04/10/2026	10,8
161P/Hartley-IRAS	09/10/2026	7,5
112P/Urata-Niijima	04/11/2026	7,2
252P/LINEAR	27/11/2026	6
2P/Encke	25/01/2027	3,7
P/2010 V1 Ikeya-Murakami	30/01/2027	11,1
73P/Schwassmann-Wachmann 3-AT	20/03/2027	11,6
73P/Schwassmann-Wachmann 3-U	25/06/2027	6,3
45P/Honda-Mrkos-Pajdusakova	30/06/2027	8,8
9P/Tempel 1	22/07/2027	10,7
73P/Schwassmann-Wachmann 3-AM	16/08/2027	10,2
73P/Schwassmann-Wachmann 3-BJ	31/08/2027	11,1
92P/Sanguin	01/09/2027	11,5
73P/Schwassmann-Wachmann 3-AB	04/09/2027	11,2
73P/Schwassmann-Wachmann 3-J	12/09/2027	11,4
73P/Schwassmann-Wachmann 3-BK	15/09/2027	11,5
73P/Schwassmann-Wachmann 3-BN	22/09/2027	11,7
73P/Schwassmann-Wachmann 3-BO	24/09/2027	11,7
P/2006 T1 (Levy)	24/09/2027	11,3
73P/Schwassmann-Wachmann 3-AX	04/10/2027	11,9
P/2011 Y1 Levy	22/11/2027	9,1
75D/Kohoutek	14/12/2027	11,5

Comet	Date	Mag
263P/Gibbs	31/01/2028	5
41P/Tuttle-Giacobini-Kresak	13/02/2028	10,8
73P/Schwassmann-Wachmann 3-AE	20/03/2028	11,6
83D/Russell 1	11/04/2028	10,7
73P/Schwassmann-Wachmann 3-W	18/04/2028	10,6
73P/Schwassmann-Wachmann 3-AI	29/04/2028	9,8
73P/Schwassmann-Wachmann 3-BS	08/05/2028	7,3
73P/Schwassmann-Wachmann 3-Q	07/06/2028	5,5
73P/Schwassmann-Wachmann 3-AL	07/08/2028	9,6
22P/Kopff	11/08/2028	8,6
73P/Schwassmann-Wachmann 3-BC	24/08/2028	10,8
73P/Schwassmann-Wachmann 3-AS	15/09/2028	11,5
73P/Schwassmann-Wachmann 3-AP	28/09/2028	11,8
P/2006 HR30 (Siding Spring)	03/12/2028	11
43P/Wolf-Harrington	04/12/2028	8,8
157P/Tritton	29/12/2028	10,6
19P/Borrelly	30/01/2029	8,4
26P/Grigg-Skjellerup	25/03/2029	11,9
81P/Wild 2	03/04/2029	10,5
118P/Shoemaker-Levy 4	06/04/2029	10,6
18D/Perrine-Mrkos	11/04/2029	9,3
46P/Wirtanen	30/10/2029	8,4
25D/Neujmin 2	06/03/2030	6,3
103P/Hartley 2	28/03/2030	11,4
2P/Encke	16/05/2030	4,2
73P/Schwassmann-Wachmann 3-AF	18/03/2031	11,7
141P/Machholz 2-D	04/04/2031	9,4
73P/Schwassmann-Wachmann 3-AW	13/04/2031	10,9
141P/Machholz 2-A	24/04/2031	3,1
167P/CINEOS	02/05/2031	11,6
73P/Schwassmann-Wachmann 3-AA	04/09/2031	11,2
210P/Christensen	14/09/2031	9,6
21P/Giacobini-Zinner	24/11/2031	8,9
144P/Kushida	04/12/2031	11,7
10P/Tempel 2	17/12/2031	10,6
49P/Arend-Rigaux	31/12/2031	11,4
P/2010 V1 Ikeya-Murakami	17/02/2032	10
5D/Brorsen	03/03/2032	-1,6
73P/Schwassmann-Wachmann 3-BR	12/03/2032	11,8
73P/Schwassmann-Wachmann 3-BF	03/04/2032	11,3
3D/Biela	12/04/2032	4,2
252P/LINEAR	16/04/2032	2,4
124P/Mrkos	20/04/2032	9,9
73P/Schwassmann-Wachmann 3-AT	11/05/2032	6,5
85P/Boethin	22/07/2032	8,6
45P/Honda-Mrkos-Pajdusakova	20/08/2032	8
73P/Schwassmann-Wachmann 3-U	29/09/2032	11,8
72D/Denning-Fujikawa	18/10/2032	-1
16P/Brooks 2	19/10/2032	10,8
P/2006 T1 (Levy)	19/12/2032	5,4
66P/du Toit	24/01/2033	9,3
9P/Tempel 1	22/02/2033	11,4
73P/Schwassmann-Wachmann 3-AV	18/03/2033	11,7
73P/Schwassmann-Wachmann 3-B	18/03/2033	11,7

Comet	Date	Mag
73P/Schwassmann-Wachmann 3-E	20/03/2033	11,6
P/2011 Y1 Levy	21/03/2033	11
73P/Schwassmann-Wachmann 3-C	23/03/2033	11,6
73P/Schwassmann-Wachmann 3-G	24/03/2033	11,5
73P/Schwassmann-Wachmann 3-K	29/03/2033	11,4
73P/Schwassmann-Wachmann 3-H	31/03/2033	11,4
73P/Schwassmann-Wachmann 3-AD	12/04/2033	10,9
73P/Schwassmann-Wachmann 3-M	14/04/2033	10,8
73P/Schwassmann-Wachmann 3-R	20/04/2033	10,5
73P/Schwassmann-Wachmann 3-AO	27/04/2033	10
73P/Schwassmann-Wachmann 3-Z	27/04/2033	10
73P/Schwassmann-Wachmann 3-AH	05/05/2033	8,6
73P/Schwassmann-Wachmann 3-AN	06/05/2033	7,9
73P/Schwassmann-Wachmann 3-L	07/05/2033	7,8
73P/Schwassmann-Wachmann 3-X	07/05/2033	7,6
263P/Gibbs	09/05/2033	8,5
112P/Urata-Niijima	03/07/2033	10,1
73P/Schwassmann-Wachmann 3-AC	14/07/2033	7,2
41P/Tuttle-Giacobini-Kresak	19/07/2033	11,7
73P/Schwassmann-Wachmann 3-P	30/07/2033	8,5
96P/Machholz 1	04/08/2033	11,4
73P/Schwassmann-Wachmann 3-AG	05/08/2033	9,4
73P/Schwassmann-Wachmann 3-AR	06/08/2033	9,5
73P/Schwassmann-Wachmann 3-AQ	08/08/2033	9,7
73P/Schwassmann-Wachmann 3-BL	21/08/2033	10,7
2P/Encke	29/08/2033	5
73P/Schwassmann-Wachmann 3-AE	10/09/2033	11,4
73P/Schwassmann-Wachmann 3-AS	30/03/2034	11,4
73P/Schwassmann-Wachmann 3-AP	14/04/2034	10,9
249P/LINEAR	23/04/2034	10
24P/Schaumasse	29/04/2034	10,1
83D/Russell 1	09/05/2034	9,5
67P/Churyumov-Gerasimenko	03/12/2034	10,2
43P/Wolf-Harrington	17/12/2034	8,8
154P/Brewington	04/02/2035	10,2
22P/Kopff	06/02/2035	10,1
46P/Wirtanen	15/04/2035	10,7
8P/Tuttle	15/04/2035	9,2
167P/CINEOS	10/06/2035	10,7
25D/Neujmin 2	20/08/2035	10
19P/Borrelly	09/12/2035	6,1
18D/Perrine-Mrkos	25/12/2035	7,6
118P/Shoemaker-Levy 4	02/01/2036	9,2
73P/Schwassmann-Wachmann 3-AW	04/04/2036	11,2
141P/Machholz 2-D	20/06/2036	9,2
141P/Machholz 2-A	13/07/2036	2,6
103P/Hartley 2	14/09/2036	9,5
73P/Schwassmann-Wachmann 3-AA	18/09/2036	11,6
162P/Siding Spring	21/10/2036	9,8
2P/Encke	13/12/2036	4,3
155P/Shoemaker 3	21/01/2037	11,6
73P/Schwassmann-Wachmann 3-V	23/03/2037	11,4
73P/Schwassmann-Wachmann 3-BD	07/04/2037	11,1
73P/Schwassmann-Wachmann 3-BG	11/04/2037	11

Comet	Date	Mag
73P/Schwassmann-Wachmann 3-BR	28/04/2037	9,9
10P/Tempel 2	09/05/2037	10
73P/Schwassmann-Wachmann 3-BF	19/05/2037	4,8
34D/Gale	20/05/2037	6
210P/Christensen	06/07/2037	11,8
252P/LINEAR	08/08/2037	6,4
5D/Brorsen	15/08/2037	0,2
73P/Schwassmann-Wachmann 3-AT	25/08/2037	10,9
239P/LINEAR	01/12/2037	11,3
45P/Honda-Mrkos-Pajdusakova	07/01/2038	7,1
62P/Tsuchinshan 1	19/01/2038	11
73P/Schwassmann-Wachmann 3-AM	01/03/2038	11,9
73P/Schwassmann-Wachmann 3-BJ	26/03/2038	11,5
P/2006 T1 (Levy)	28/03/2038	11,1
73P/Schwassmann-Wachmann 3-AB	02/04/2038	11,3
73P/Schwassmann-Wachmann 3-J	13/04/2038	10,9
73P/Schwassmann-Wachmann 3-BK	17/04/2038	10,7
73P/Schwassmann-Wachmann 3-BN	25/04/2038	10,2
73P/Schwassmann-Wachmann 3-BO	28/04/2038	10
73P/Schwassmann-Wachmann 3-AX	04/05/2038	8,6
124P/Mrkos	09/05/2038	10,4
73P/Schwassmann-Wachmann 3-N	14/05/2038	5,9
73P/Schwassmann-Wachmann 3-BQ	31/05/2038	4,9
21P/Giacobini-Zinner	25/06/2038	10,1
73P/Schwassmann-Wachmann 3	25/06/2038	6,4
73P/Schwassmann-Wachmann 3-AU	27/07/2038	8,1
73P/Schwassmann-Wachmann 3-AK	28/07/2038	8,3
9P/Tempel 1	10/08/2038	11,1
73P/Schwassmann-Wachmann 3-B	12/08/2038	10,1
73P/Schwassmann-Wachmann 3-AV	13/08/2038	10,1
73P/Schwassmann-Wachmann 3-E	15/08/2038	10,2
73P/Schwassmann-Wachmann 3-C	18/08/2038	10,5
73P/Schwassmann-Wachmann 3-G	19/08/2038	10,5
263P/Gibbs	20/08/2038	9,2
73P/Schwassmann-Wachmann 3-K	23/08/2038	10,7
73P/Schwassmann-Wachmann 3-H	26/08/2038	10,9
73P/Schwassmann-Wachmann 3-AD	10/09/2038	11,4
73P/Schwassmann-Wachmann 3-M	12/09/2038	11,5
73P/Schwassmann-Wachmann 3-R	21/09/2038	11,7
73P/Schwassmann-Wachmann 3-AO	01/10/2038	11,9
73P/Schwassmann-Wachmann 3-Z	02/10/2038	11,9
96P/Machholz 1	19/11/2038	8,6
3D/Biela	24/11/2038	2,7
73P/Schwassmann-Wachmann 3-W	02/04/2039	11,3
73P/Schwassmann-Wachmann 3-AI	21/04/2039	10,5
73P/Schwassmann-Wachmann 3-BS	08/05/2039	7,7
73P/Schwassmann-Wachmann 3-Q	22/06/2039	6,2
167P/CINEOS	12/07/2039	10,2
73P/Schwassmann-Wachmann 3-AL	24/08/2039	10,8
73P/Schwassmann-Wachmann 3-BC	26/09/2039	11,8
112P/Urata-Niijima	18/02/2040	9,8
2P/Encke	03/04/2040	3,7
83D/Russell 1	13/05/2040	9,2
7P/Pons-Winnecke	09/07/2040	10,2

Comet	Date	Mag
46P/Wirtanen	18/09/2040	10
78P/Gehrels 2	01/11/2040	11,7

COMETE VISIBILI AD OCCHIO NUDO
NAKED EYES COMETS

Cometa-comet		Date	Mag
2P/Encke		20/11/2013	4,6
C/2012 S1 ISON		28/11/2013	-5,6
72D/Denning-Fujikawa		11/10/2014	-0,9
C/2012 K1 PANSTARRS		14/11/2014	6
34D/Gale		26/05/2015	5,7
141P/Machholz 2-A		11/08/2015	2
5D/Brorsen		09/10/2015	-0,3
18D/Perrine-Mrkos		11/11/2015	5
252P/LINEAR	*	07/04/2016	-0,7
2P/Encke		09/03/2017	3,5
73P/Schwassmann-Wachmann 3-Q	*	24/05/2017	4,5
3D/Biela		27/11/2018	0
46P/Wirtanen		27/12/2018	5,5
2P/Encke		26/06/2020	4,7
141P/Machholz 2-D		24/10/2020	5,4
141P/Machholz 2-A		15/11/2020	-1,1
5D/Brorsen		30/03/2021	-0,8
73P/Schwassmann-Wachmann 3-N		15/05/2022	5,7
73P/Schwassmann-Wachmann 3-BQ		22/05/2022	4,5
73P/Schwassmann-Wachmann 3		23/05/2022	4,6
2P/Encke		10/10/2023	4,9
72D/Denning-Fujikawa		15/10/2023	-1
3D/Biela		16/08/2025	4,9
141P/Machholz 2-A		01/02/2026	2,4
34D/Gale		23/05/2026	5,9
5D/Brorsen		12/09/2026	0
252P/LINEAR		27/11/2026	6
2P/Encke		25/01/2027	3,7
263P/Gibbs	*	31/01/2028	5
73P/Schwassmann-Wachmann 3-Q		07/06/2028	5,5
2P/Encke		16/05/2030	4,2
141P/Machholz 2-A		24/04/2031	3,1
5D/Brorsen		03/03/2032	-1,6
3D/Biela		12/04/2032	4,2
252P/LINEAR		16/04/2032	2,4
72D/Denning-Fujikawa		18/10/2032	-1
P/2006 T1 (Levy)		19/12/2032	5,4
2P/Encke		29/08/2033	5
141P/Machholz 2-A		13/07/2036	2,6
2P/Encke		13/12/2036	4,3
73P/Schwassmann-Wachmann 3-BF		19/05/2037	4,8
34D/Gale		20/05/2037	6
5D/Brorsen		15/08/2037	0,2
73P/Schwassmann-Wachmann 3-N		14/05/2038	5,9
73P/Schwassmann-Wachmann 3-BQ		31/05/2038	4,9
3D/Biela		24/11/2038	2,7
2P/Encke		03/04/2040	3,7

* Dati incerti

EFFEMERIDI - EPHEMERIDES

Cometa 2P/Encke

Data	AR [h m s]	Dec [° ']	Alt. [°]	Az. [°]	Elongazione [°]
2013:10:18	10:27:02	+32°36'	4.0	48.4	61.0° W
2013:10:19	10:38:57	+31°03'	1.7	48.2	58.9° W
2013:10:20	10:50:27	+29°26'	-0.6	48.0	56.9° W
2013:10:21	11:01:30	+27°46'	-2.9	47.8	54.8° W
2013:10:22	11:12:06	+26°03'	-5.2	47.7	52.8° W
2013:10:23	11:22:16	+24°20'	-7.4	47.6	50.9° W
2013:10:24	11:31:58	+22°35'	-9.6	47.6	49.0° W
2013:10:25	11:41:15	+20°51'	-11.7	47.6	47.1° W
2013:10:26	11:50:07	+19°07'	-13.8	47.7	45.3° W
2013:10:27	11:58:35	+17°24'	-15.8	47.8	43.6° W
2013:10:28	12:06:41	+15°42'	-17.8	48.0	41.9° W
2013:10:29	12:14:25	+14°02'	-19.7	48.2	40.3° W
2013:10:30	12:21:50	+12°25'	-21.5	48.5	38.8° W
2013:10:31	12:28:57	+10°49'	-23.2	48.8	37.4° W
2013:11:01	12:35:47	+09°15'	-24.9	49.1	36.0° W
2013:11:02	12:42:22	+07°44'	-26.5	49.5	34.8° W
2013:11:03	12:48:43	+06°16'	-28.1	49.9	33.5° W
2013:11:04	12:54:52	+04°49'	-29.6	50.3	32.4° W
2013:11:05	13:00:51	+03°25'	-31.0	50.8	31.3° W
2013:11:06	13:06:40	+02°04'	-32.4	51.2	30.3° W
2013:11:07	13:12:23	+00°44'	-33.7	51.7	29.3° W
2013:11:08	13:17:59	-00°34'	-35.0	52.2	28.4° W
2013:11:09	13:23:31	-01°50'	-36.3	52.8	27.5° W
2013:11:10	13:29:00	-03°04'	-37.5	53.3	26.6° W
2013:11:11	13:34:28	-04°17'	-38.7	53.8	25.8° W
2013:11:12	13:39:57	-05°28'	-39.9	54.4	25.0° W
2013:11:13	13:45:28	-06°37'	-41.0	54.9	24.2° W
2013:11:14	13:51:03	-07°46'	-42.2	55.4	23.5° W
2013:11:15	13:56:44	-08°53'	-43.3	55.9	22.7° W
2013:11:16	14:02:31	-10°00'	-44.5	56.4	21.9° W
2013:11:17	14:08:27	-11°05'	-45.6	56.8	21.1° W
2013:11:18	14:14:33	-12°09'	-46.8	57.2	20.3° W
2013:11:19	14:20:49	-13°11'	-48.0	57.5	19.5° W
2013:11:20	14:27:16	-14°13'	-49.1	57.8	18.7° W
2013:11:21	14:33:53	-15°13'	-50.3	58.1	17.9° W
2013:11:22	14:40:41	-16°11'	-51.5	58.3	17.1° W
2013:11:23	14:47:38	-17°07'	-52.7	58.4	16.2° W
2013:11:24	14:54:43	-18°01'	-53.8	58.5	15.4° W
2013:11:25	15:01:54	-18°52'	-55.0	58.5	14.6° W
2013:11:26	15:09:10	-19°42'	-56.1	58.4	13.7° W
2013:11:27	15:16:28	-20°28'	-57.2	58.3	13.0° W
2013:11:28	15:23:47	-21°12'	-58.3	58.2	12.2° W
2013:11:29	15:31:05	-21°53'	-59.3	57.9	11.5° W
2013:11:30	15:38:22	-22°31'	-60.3	57.7	10.8° W
2013:12:01	15:45:35	-23°07'	-61.3	57.4	10.1° W
2013:12:02	15:52:45	-23°40'	-62.2	57.0	9.5° W
2013:12:03	15:59:50	-24°11'	-63.0	56.7	8.9° W
2013:12:04	16:06:49	-24°39'	-63.9	56.2	8.4° W
2013:12:05	16:13:42	-25°05'	-64.6	55.8	7.9° W
2013:12:06	16:20:29	-25°29'	-65.4	55.3	7.5° W

Data	AR [h m s]	Dec [° ']	Alt. [°]	Az. [°]	Elongazione [°]
2013:12:07	16:27:10	-25°51'	-66.0	54.8	7.1° W
2013:12:08	16:33:44	-26°11'	-66.7	54.3	6.8° W
2013:12:09	16:40:11	-26°29'	-67.3	53.7	6.5° W
2013:12:10	16:46:32	-26°45'	-67.9	53.2	6.2° W

Data	Ora	Dist(Sol) [AU]	Dist(Terra) [AU]	Magnit. [mag]
2013:10:18	00:00	0.871	0.478	9.0
2013:10:19	00:00	0.854	0.480	8.9
2013:10:20	00:00	0.836	0.482	8.8
2013:10:21	00:00	0.819	0.486	8.6
2013:10:22	00:00	0.801	0.491	8.5
2013:10:23	00:00	0.783	0.497	8.4
2013:10:24	00:00	0.765	0.504	8.3
2013:10:25	00:00	0.747	0.513	8.1
2013:10:26	00:00	0.729	0.522	8.0
2013:10:27	00:00	0.710	0.533	7.9
2013:10:28	00:00	0.692	0.544	7.8
2013:10:29	00:00	0.673	0.557	7.7
2013:10:30	00:00	0.655	0.570	7.5
2013:10:31	00:00	0.637	0.585	7.4
2013:11:01	00:00	0.618	0.600	7.3

Data	Ora	Dist(Sol) [AU]	Dist(Terra) [AU]	Magnit. [mag]
2013:11:02	00:00	0.600	0.617	7.1
2013:11:03	00:00	0.581	0.634	7.0
2013:11:04	00:00	0.563	0.652	6.8
2013:11:05	00:00	0.544	0.672	6.7
2013:11:06	00:00	0.526	0.692	6.5
2013:11:07	00:00	0.508	0.713	6.4
2013:11:08	00:00	0.490	0.734	6.2
2013:11:09	00:00	0.473	0.757	6.0
2013:11:10	00:00	0.456	0.781	5.8
2013:11:11	00:00	0.440	0.805	5.7
2013:11:12	00:00	0.424	0.830	5.5
2013:11:13	00:00	0.409	0.857	5.3
2013:11:14	00:00	0.395	0.883	5.2
2013:11:15	00:00	0.382	0.911	5.0
2013:11:16	00:00	0.370	0.939	4.9
2013:11:17	00:00	0.359	0.968	4.8
2013:11:18	00:00	0.351	0.997	4.7
2013:11:19	00:00	0.344	1.026	4.6
2013:11:20	00:00	0.339	1.056	4.6
2013:11:21	00:00	0.337	1.085	4.6
2013:11:22	00:00	0.336	1.115	4.6
2013:11:23	00:00	0.338	1.143	4.7
2013:11:24	00:00	0.342	1.172	4.9
2013:11:25	00:00	0.348	1.199	5.0
2013:11:26	00:00	0.356	1.226	5.2
2013:11:27	00:00	0.366	1.253	5.4
2013:11:28	00:00	0.377	1.278	5.7
2013:11:29	00:00	0.389	1.303	5.9
2013:11:30	00:00	0.403	1.327	6.2
2013:12:01	00:00	0.418	1.351	6.5
2013:12:02	00:00	0.433	1.374	6.7
2013:12:03	00:00	0.450	1.397	7.0
2013:12:04	00:00	0.466	1.419	7.3
2013:12:05	00:00	0.484	1.440	7.6
2013:12:06	00:00	0.501	1.461	7.8
2013:12:07	00:00	0.519	1.482	8.1
2013:12:08	00:00	0.537	1.503	8.3
2013:12:09	00:00	0.555	1.523	8.6
2013:12:10	00:00	0.574	1.543	8.8

Cometa C/2012 S1 ISON

Data	AR [h m s]	Dec [° ']	Alt. [°]	Az. [°]	Elongazione [°]
2013:10:11	09:58:44	+15°11'	-9.0	60.1	51.4° W
2013:10:12	10:01:21	+14°55'	-9.0	60.6	51.7° W
2013:10:13	10:04:01	+14°37'	-9.0	61.0	52.0° W
2013:10:14	10:06:45	+14°19'	-9.0	61.4	52.2° W
2013:10:15	10:09:34	+14°01'	-9.0	61.8	52.5° W
2013:10:16	10:12:26	+13°42'	-9.1	62.2	52.7° W
2013:10:17	10:15:23	+13°22'	-9.2	62.6	52.9° W
2013:10:18	10:18:26	+13°01'	-9.3	63.0	53.1° W
2013:10:19	10:21:33	+12°40'	-9.4	63.4	53.2° W
2013:10:20	10:24:46	+12°17'	-9.6	63.8	53.3° W
2013:10:21	10:28:04	+11°54'	-9.7	64.2	53.4° W
2013:10:22	10:31:29	+11°30'	-10.0	64.6	53.5° W
2013:10:23	10:35:01	+11°04'	-10.2	65.0	53.5° W
2013:10:24	10:38:40	+10°38'	-10.5	65.3	53.5° W
2013:10:25	10:42:26	+10°11'	-10.8	65.7	53.5° W
2013:10:26	10:46:20	+09°42'	-11.1	66.0	53.4° W
2013:10:27	10:50:22	+09°12'	-11.5	66.4	53.3° W
2013:10:28	10:54:34	+08°40'	-11.9	66.7	53.1° W
2013:10:29	10:58:56	+08°07'	-12.4	67.0	52.9° W
2013:10:30	11:03:28	+07°33'	-12.9	67.3	52.7° W
2013:10:31	11:08:11	+06°56'	-13.5	67.6	52.4° W
2013:11:01	11:13:07	+06°18'	-14.1	67.9	52.0° W
2013:11:02	11:18:15	+05°38'	-14.8	68.1	51.5° W
2013:11:03	11:23:37	+04°56'	-15.5	68.3	51.0° W
2013:11:04	11:29:13	+04°12'	-16.3	68.6	50.5° W
2013:11:05	11:35:06	+03°25'	-17.2	68.8	49.8° W
2013:11:06	11:41:15	+02°36'	-18.2	69.0	49.1° W
2013:11:07	11:47:43	+01°44'	-19.3	69.1	48.3° W
2013:11:08	11:54:30	+00°50'	-20.4	69.2	47.3° W
2013:11:09	12:01:37	-00°07'	-21.6	69.3	46.3° W
2013:11:10	12:09:07	-01°07'	-23.0	69.4	45.2° W
2013:11:11	12:17:01	-02°10'	-24.4	69.4	44.0° W
2013:11:12	12:25:19	-03°16'	-26.0	69.4	42.7° W
2013:11:13	12:34:04	-04°25'	-27.6	69.4	41.2° W
2013:11:14	12:43:18	-05°37'	-29.4	69.2	39.6° W
2013:11:15	12:53:01	-06°52'	-31.3	69.1	37.9° W
2013:11:16	13:03:15	-08°09'	-33.3	68.8	36.1° W
2013:11:17	13:14:02	-09°28'	-35.4	68.5	34.2° W
2013:11:18	13:25:22	-10°49'	-37.7	68.0	32.1° W
2013:11:19	13:37:16	-12°11'	-40.0	67.5	29.9° W
2013:11:20	13:49:46	-13°34'	-42.5	66.8	27.6° W
2013:11:21	14:02:52	-14°57'	-45.0	66.0	25.2° W
2013:11:22	14:16:34	-16°18'	-47.6	64.9	22.6° W
2013:11:23	14:30:54	-17°37'	-50.3	63.7	20.0° W
2013:11:24	14:45:53	-18°53'	-53.0	62.0	17.3° W
2013:11:25	15:01:35	-20°05'	-55.8	60.0	14.5° W
2013:11:26	15:18:10	-21°10'	-58.6	57.4	11.5° W
2013:11:27	15:35:59	-22°05'	-61.5	53.7	8.4° W
2013:11:28	15:56:17	-22°45'	-64.3	48.2	4.8° W
2013:11:29	16:24:17	-20°16'	-65.2	33.9	1.5° E

Data	AR [h m s]	Dec [° ']	Alt. [°]	Az. [°]	Elongazione [°]
2013:11:30	16:22:44	-16°33'	-61.3	33.0	5.1° W
2013:12:01	16:20:36	-14°10'	-58.5	33.6	7.9° W
2013:12:02	16:18:50	-12°07'	-56.0	34.3	10.4° W
2013:12:03	16:17:22	-10°12'	-53.7	34.9	12.8° W
2013:12:04	16:16:08	-08°20'	-51.5	35.5	15.2° W
2013:12:05	16:15:08	-06°30'	-49.3	36.0	17.6° W
2013:12:06	16:14:17	-04°39'	-47.1	36.3	19.9° W
2013:12:07	16:13:37	-02°47'	-44.9	36.6	22.2° W
2013:12:08	16:13:04	-00°53'	-42.7	36.9	24.6° W
2013:12:09	16:12:39	+01°05'	-40.5	37.0	26.9° W
2013:12:10	16:12:22	+03°06'	-38.2	37.0	29.4° W
2013:12:11	16:12:11	+05°12'	-35.9	37.0	31.8° W
2013:12:12	16:12:07	+07°23'	-33.5	36.9	34.3° W
2013:12:13	16:12:08	+09°39'	-31.0	36.7	36.9° W
2013:12:14	16:12:16	+12°02'	-28.5	36.4	39.6° W
2013:12:15	16:12:30	+14°30'	-25.8	36.1	42.3° W
2013:12:16	16:12:51	+17°06'	-23.1	35.6	45.1° W
2013:12:17	16:13:17	+19°50'	-20.3	35.1	48.0° W
2013:12:18	16:13:50	+22°40'	-17.4	34.5	51.0° W
2013:12:19	16:14:30	+25°38'	-14.5	33.7	54.0° W
2013:12:20	16:15:17	+28°44'	-11.4	32.9	57.2° W
2013:12:21	16:16:13	+31°57'	-8.2	32.0	60.4° W
2013:12:22	16:17:17	+35°17'	-5.0	31.0	63.7° W
2013:12:23	16:18:31	+38°42'	-1.7	29.8	67.0° W
2013:12:24	16:19:56	+42°14'	1.6	28.6	70.4° W
2013:12:25	16:21:35	+45°49'	4.9	27.2	73.8° W
2013:12:26	16:23:30	+49°27'	8.3	25.7	77.2° W
2013:12:27	16:25:44	+53°07'	11.6	24.1	80.6° W
2013:12:28	16:28:22	+56°46'	14.8	22.3	83.9° W
2013:12:29	16:31:32	+60°24'	17.9	20.4	87.2° W
2013:12:30	16:35:23	+63°58'	21.0	18.4	90.3° W
2013:12:31	16:40:12	+67°27'	23.9	16.3	93.3° W
2014:01:01	16:46:23	+70°50'	26.6	14.1	96.2° W
2014:01:02	16:54:38	+74°06'	29.1	11.7	98.9° W
2014:01:03	17:06:17	+77°13'	31.5	9.3	101.4° W
2014:01:04	17:23:58	+80°09'	33.6	6.8	103.7° W
2014:01:05	17:53:48	+82°52'	35.5	4.2	105.8° W
2014:01:06	18:52:01	+85°13'	37.3	1.6	107.8° W
2014:01:07	20:58:05	+86°46'	38.8	358.9	109.5° E
2014:01:08	23:55:21	+86°37'	40.1	356.2	111.1° E
2014:01:09	01:39:37	+85°08'	41.2	353.6	112.5° E
2014:01:10	02:28:02	+83°16'	42.1	350.9	113.7° E
2014:01:11	02:53:56	+81°21'	42.9	348.4	114.7° E
2014:01:12	03:09:52	+79°30'	43.5	345.9	115.6° E
2014:01:13	03:20:41	+77°43'	44.0	343.4	116.4° E
2014:01:14	03:28:34	+76°01'	44.3	341.1	117.0° E
2014:01:15	03:34:38	+74°25'	44.5	338.9	117.5° E
2014:01:16	03:39:30	+72°54'	44.6	336.7	117.9° E
2014:01:17	03:43:32	+71°28'	44.6	334.7	118.2° E
2014:01:18	03:46:58	+70°07'	44.5	332.8	118.3° E
2014:01:19	03:49:57	+68°50'	44.4	331.0	118.4° E
2014:01:20	03:52:36	+67°37'	44.2	329.4	118.5° E

Data	AR [h m s]	Dec [° ']	Alt. [°]	Az. [°]	Elongazione [°]
2014:01:21	03:54:59	+66°28'	43.9	327.8	118.4° E
2014:01:22	03:57:10	+65°23'	43.6	326.3	118.3° E
2014:01:23	03:59:12	+64°21'	43.3	325.0	118.1° E
2014:01:24	04:01:05	+63°23'	42.9	323.7	117.9° E
2014:01:25	04:02:51	+62°27'	42.5	322.5	117.6° E
2014:01:26	04:04:33	+61°35'	42.0	321.5	117.3° E
2014:01:27	04:06:09	+60°45'	41.6	320.5	116.9° E
2014:01:28	04:07:42	+59°57'	41.1	319.5	116.5° E
2014:01:29	04:09:12	+59°12'	40.6	318.7	116.0° E
2014:01:30	04:10:39	+58°29'	40.0	317.9	115.6° E
2014:01:31	04:12:04	+57°47'	39.5	317.2	115.1° E
2014:02:01	04:13:27	+57°08'	39.0	316.5	114.6° E
2014:02:02	04:14:48	+56°31'	38.4	315.9	114.0° E
2014:02:03	04:16:08	+55°55'	37.9	315.4	113.5° E
2014:02:04	04:17:27	+55°21'	37.3	314.9	112.9° E
2014:02:05	04:18:44	+54°48'	36.8	314.4	112.3° E

Data	Ora	Dist(Sol) [AU]	Dist(Terra) [AU]	Magnit. [mag]
2013:10:11	00:00	1.457	1.854	9.0
2013:10:12	00:00	1.437	1.824	8.9
2013:10:13	00:00	1.417	1.794	8.8
2013:10:14	00:00	1.397	1.764	8.7
2013:10:15	00:00	1.376	1.734	8.6
2013:10:16	00:00	1.355	1.703	8.5
2013:10:17	00:00	1.334	1.673	8.4
2013:10:18	00:00	1.313	1.643	8.3
2013:10:19	00:00	1.292	1.613	8.2
2013:10:20	00:00	1.271	1.583	8.0
2013:10:21	00:00	1.249	1.553	7.9
2013:10:22	00:00	1.228	1.523	7.8
2013:10:23	00:00	1.206	1.493	7.7
2013:10:24	00:00	1.183	1.463	7.6
2013:10:25	00:00	1.161	1.434	7.4
2013:10:26	00:00	1.139	1.404	7.3
2013:10:27	00:00	1.116	1.375	7.2
2013:10:28	00:00	1.093	1.346	7.0
2013:10:29	00:00	1.069	1.317	6.9
2013:10:30	00:00	1.046	1.288	6.7
2013:10:31	00:00	1.022	1.260	6.6
2013:11:01	00:00	0.998	1.232	6.4
2013:11:02	00:00	0.974	1.204	6.3
2013:11:03	00:00	0.949	1.177	6.1
2013:11:04	00:00	0.924	1.150	6.0
2013:11:05	00:00	0.899	1.124	5.8
2013:11:06	00:00	0.873	1.098	5.6
2013:11:07	00:00	0.847	1.073	5.4

Data	Ora	Dist(Sol) [AU]	Dist(Terra) [AU]	Magnit. [mag]
2013:11:08	00:00	0.821	1.049	5.2
2013:11:09	00:00	0.794	1.026	5.1
2013:11:10	00:00	0.766	1.003	4.9
2013:11:11	00:00	0.739	0.982	4.6
2013:11:12	00:00	0.710	0.962	4.4
2013:11:13	00:00	0.681	0.943	4.2
2013:11:14	00:00	0.652	0.925	4.0
2013:11:15	00:00	0.622	0.909	3.7
2013:11:16	00:00	0.591	0.895	3.5
2013:11:17	00:00	0.559	0.883	3.2
2013:11:18	00:00	0.526	0.872	2.9
2013:11:19	00:00	0.493	0.865	2.6
2013:11:20	00:00	0.458	0.859	2.3
2013:11:21	00:00	0.422	0.857	1.9
2013:11:22	00:00	0.384	0.857	1.5
2013:11:23	00:00	0.344	0.862	1.0
2013:11:24	00:00	0.303	0.869	0.5
2013:11:25	00:00	0.258	0.882	-0.2
2013:11:26	00:00	0.209	0.899	-1.0
2013:11:27	00:00	0.153	0.923	-2.3
2013:11:28	00:00	0.087	0.958	-4.7
2013:11:29	00:00	0.026	0.981	-9.8
2013:11:30	00:00	0.111	0.915	-3.8
2013:12:01	00:00	0.173	0.869	-1.9
2013:12:02	00:00	0.225	0.832	-0.9
2013:12:03	00:00	0.273	0.799	-0.1
2013:12:04	00:00	0.317	0.768	0.4
2013:12:05	00:00	0.358	0.740	0.9
2013:12:06	00:00	0.397	0.714	1.3
2013:12:07	00:00	0.434	0.689	1.6
2013:12:08	00:00	0.470	0.666	1.8
2013:12:09	00:00	0.504	0.644	2.1
2013:12:10	00:00	0.537	0.622	2.3
2013:12:11	00:00	0.570	0.602	2.5
2013:12:12	00:00	0.601	0.583	2.6
2013:12:13	00:00	0.632	0.564	2.8
2013:12:14	00:00	0.662	0.547	2.9
2013:12:15	00:00	0.691	0.531	3.0
2013:12:16	00:00	0.720	0.515	3.1
2013:12:17	00:00	0.748	0.501	3.2
2013:12:18	00:00	0.776	0.487	3.3
2013:12:19	00:00	0.803	0.475	3.4
2013:12:20	00:00	0.830	0.464	3.5
2013:12:21	00:00	0.856	0.454	3.6
2013:12:22	00:00	0.882	0.446	3.7
2013:12:23	00:00	0.907	0.439	3.8
2013:12:24	00:00	0.932	0.434	3.9
2013:12:25	00:00	0.957	0.430	4.0
2013:12:26	00:00	0.982	0.428	4.1
2013:12:27	00:00	1.006	0.427	4.2
2013:12:28	00:00	1.030	0.428	4.3
2013:12:29	00:00	1.054	0.430	4.4

Data	Ora	Dist(Sol) [AU]	Dist(Terra) [AU]	Magnit. [mag]
2013:12:30	00:00	1.077	0.435	4.5
2013:12:31	00:00	1.100	0.440	4.6
2014:01:01	00:00	1.123	0.448	4.8
2014:01:02	00:00	1.146	0.456	4.9
2014:01:03	00:00	1.169	0.467	5.0
2014:01:04	00:00	1.191	0.478	5.2
2014:01:05	00:00	1.213	0.491	5.3
2014:01:06	00:00	1.235	0.505	5.4
2014:01:07	00:00	1.256	0.520	5.6
2014:01:08	00:00	1.278	0.536	5.7
2014:01:09	00:00	1.299	0.553	5.9
2014:01:10	00:00	1.320	0.571	6.0
2014:01:11	00:00	1.341	0.589	6.1
2014:01:12	00:00	1.362	0.609	6.3
2014:01:13	00:00	1.383	0.629	6.4
2014:01:14	00:00	1.403	0.650	6.5
2014:01:15	00:00	1.424	0.671	6.7
2014:01:16	00:00	1.444	0.693	6.8
2014:01:17	00:00	1.464	0.715	6.9
2014:01:18	00:00	1.484	0.738	7.1
2014:01:19	00:00	1.504	0.762	7.2
2014:01:20	00:00	1.524	0.785	7.3
2014:01:21	00:00	1.543	0.809	7.4
2014:01:22	00:00	1.563	0.834	7.5
2014:01:23	00:00	1.582	0.859	7.7
2014:01:24	00:00	1.601	0.884	7.8
2014:01:25	00:00	1.620	0.909	7.9
2014:01:26	00:00	1.639	0.935	8.0
2014:01:27	00:00	1.658	0.961	8.1
2014:01:28	00:00	1.677	0.987	8.2
2014:01:29	00:00	1.696	1.014	8.3
2014:01:30	00:00	1.714	1.041	8.4
2014:01:31	00:00	1.733	1.068	8.5
2014:02:01	00:00	1.751	1.095	8.6
2014:02:02	00:00	1.769	1.122	8.7
2014:02:03	00:00	1.787	1.149	8.8
2014:02:04	00:00	1.805	1.177	8.9
2014:02:05	00:00	1.823	1.205	9.0

Cometa C/2012 K1 PANSTARRS

Data	Ora	AR [h m s]	Dec [° ']	Elongazione [°]
2014:07:01	00:00	10:45:01	+31°50'	54.4° E
2014:07:02	00:00	10:43:50	+31°31'	53.4° E
2014:07:03	00:00	10:42:42	+31°11'	52.3° E
2014:07:04	00:00	10:41:36	+30°52'	51.3° E
2014:07:05	00:00	10:40:32	+30°33'	50.2° E
2014:07:06	00:00	10:39:30	+30°14'	49.2° E
2014:07:07	00:00	10:38:31	+29°55'	48.1° E
2014:07:08	00:00	10:37:34	+29°36'	47.1° E
2014:07:09	00:00	10:36:38	+29°17'	46.1° E
2014:07:10	00:00	10:35:45	+28°59'	45.0° E
2014:07:11	00:00	10:34:53	+28°40'	44.0° E
2014:07:12	00:00	10:34:03	+28°22'	43.0° E
2014:07:13	00:00	10:33:14	+28°03'	42.0° E
2014:07:14	00:00	10:32:27	+27°45'	41.0° E
2014:07:15	00:00	10:31:42	+27°27'	40.0° E
2014:07:16	00:00	10:30:58	+27°09'	39.0° E
2014:07:17	00:00	10:30:16	+26°51'	38.0° E
2014:07:18	00:00	10:29:35	+26°33'	37.0° E
2014:07:19	00:00	10:28:55	+26°15'	36.0° E
2014:07:20	00:00	10:28:16	+25°57'	35.0° E
2014:07:21	00:00	10:27:39	+25°40'	34.0° E
2014:07:22	00:00	10:27:03	+25°22'	33.0° E
2014:07:23	00:00	10:26:27	+25°04'	32.0° E
2014:07:24	00:00	10:25:53	+24°47'	31.0° E
2014:07:25	00:00	10:25:20	+24°30'	30.0° E
2014:07:26	00:00	10:24:48	+24°12'	29.0° E
2014:07:27	00:00	10:24:16	+23°55'	28.1° E
2014:07:28	00:00	10:23:46	+23°38'	27.1° E
2014:07:29	00:00	10:23:16	+23°21'	26.1° E
2014:07:30	00:00	10:22:47	+23°03'	25.2° E
2014:07:31	00:00	10:22:19	+22°46'	24.2° E
2014:08:01	00:00	10:21:52	+22°29'	23.2° E
2014:08:02	00:00	10:21:25	+22°12'	22.3° E
2014:08:03	00:00	10:20:59	+21°55'	21.3° E
2014:08:04	00:00	10:20:33	+21°38'	20.3° E
2014:08:05	00:00	10:20:08	+21°21'	19.4° E
2014:08:06	00:00	10:19:43	+21°05'	18.4° E
2014:08:07	00:00	10:19:19	+20°48'	17.5° E
2014:08:08	00:00	10:18:55	+20°31'	16.6° E
2014:08:09	00:00	10:18:32	+20°14'	15.6° E
2014:08:10	00:00	10:18:09	+19°57'	14.7° E
2014:08:11	00:00	10:17:47	+19°40'	13.8° E
2014:08:12	00:00	10:17:25	+19°24'	12.9° E
2014:08:13	00:00	10:17:03	+19°07'	12.0° E
2014:08:14	00:00	10:16:41	+18°50'	11.1° E
2014:08:15	00:00	10:16:20	+18°33'	10.2° E
2014:08:16	00:00	10:15:58	+18°16'	9.4° E
2014:08:17	00:00	10:15:37	+17°59'	8.5° E
2014:08:18	00:00	10:15:16	+17°42'	7.8° E
2014:08:19	00:00	10:14:55	+17°25'	7.0° E

Data	Ora	AR [h m s]	Dec [° ']	Elongazione [°]
2014:08:20	00:00	10:14:35	+17°08'	6.3° E
2014:08:21	00:00	10:14:14	+16°51'	5.7° E
2014:08:22	00:00	10:13:53	+16°34'	5.3° E
2014:08:23	00:00	10:13:33	+16°17'	5.0° E
2014:08:24	00:00	10:13:12	+16°00'	4.8° E
2014:08:25	00:00	10:12:52	+15°43'	4.9° W
2014:08:26	00:00	10:12:31	+15°25'	5.1° W
2014:08:27	00:00	10:12:10	+15°08'	5.5° W
2014:08:28	00:00	10:11:49	+14°50'	6.1° W
2014:08:29	00:00	10:11:28	+14°32'	6.7° W
2014:08:30	00:00	10:11:07	+14°15'	7.5° W
2014:08:31	00:00	10:10:46	+13°57'	8.3° W
2014:09:01	00:00	10:10:24	+13°39'	9.1° W
2014:09:02	00:00	10:10:02	+13°21'	9.9° W
2014:09:03	00:00	10:09:40	+13°03'	10.8° W
2014:09:04	00:00	10:09:17	+12°44'	11.7° W
2014:09:05	00:00	10:08:55	+12°26'	12.6° W
2014:09:06	00:00	10:08:32	+12°07'	13.5° W
2014:09:07	00:00	10:08:08	+11°48'	14.5° W
2014:09:08	00:00	10:07:44	+11°29'	15.4° W
2014:09:09	00:00	10:07:20	+11°10'	16.4° W
2014:09:10	00:00	10:06:55	+10°51'	17.4° W
2014:09:11	00:00	10:06:30	+10°31'	18.3° W
2014:09:12	00:00	10:06:04	+10°11'	19.3° W
2014:09:13	00:00	10:05:37	+09°51'	20.3° W
2014:09:14	00:00	10:05:10	+09°31'	21.3° W
2014:09:15	00:00	10:04:43	+09°10'	22.3° W
2014:09:16	00:00	10:04:14	+08°50'	23.3° W
2014:09:17	00:00	10:03:45	+08°28'	24.3° W
2014:09:18	00:00	10:03:16	+08°07'	25.3° W
2014:09:19	00:00	10:02:45	+07°46'	26.3° W
2014:09:20	00:00	10:02:14	+07°24'	27.3° W
2014:09:21	00:00	10:01:41	+07°01'	28.3° W
2014:09:22	00:00	10:01:08	+06°39'	29.3° W
2014:09:23	00:00	10:00:34	+06°16'	30.4° W
2014:09:24	00:00	09:59:59	+05°52'	31.4° W
2014:09:25	00:00	09:59:23	+05°29'	32.4° W
2014:09:26	00:00	09:58:45	+05°04'	33.5° W
2014:09:27	00:00	09:58:07	+04°40'	34.5° W
2014:09:28	00:00	09:57:27	+04°15'	35.6° W
2014:09:29	00:00	09:56:45	+03°49'	36.6° W
2014:09:30	00:00	09:56:03	+03°23'	37.7° W
2014:10:01	00:00	09:55:19	+02°57'	38.7° W
2014:10:02	00:00	09:54:33	+02°30'	39.8° W
2014:10:03	00:00	09:53:46	+02°02'	40.9° W
2014:10:04	00:00	09:52:57	+01°34'	42.0° W
2014:10:05	00:00	09:52:06	+01°05'	43.1° W
2014:10:06	00:00	09:51:13	+00°35'	44.2° W
2014:10:07	00:00	09:50:18	+00°05'	45.3° W
2014:10:08	00:00	09:49:21	-00°26'	46.4° W
2014:10:09	00:00	09:48:22	-00°57'	47.5° W
2014:10:10	00:00	09:47:20	-01°30'	48.6° W

Data	Ora	AR [h m s]	Dec [° ']	Elongazione [°]
2014:10:11	00:00	09:46:16	-02°03'	49.7° W
2014:10:12	00:00	09:45:09	-02°37'	50.9° W
2014:10:13	00:00	09:43:59	-03°12'	52.0° W
2014:10:14	00:00	09:42:46	-03°48'	53.2° W
2014:10:15	00:00	09:41:30	-04°25'	54.3° W
2014:10:16	00:00	09:40:10	-05°03'	55.5° W
2014:10:17	00:00	09:38:47	-05°42'	56.7° W
2014:10:18	00:00	09:37:20	-06°22'	57.9° W
2014:10:19	00:00	09:35:48	-07°03'	59.1° W
2014:10:20	00:00	09:34:13	-07°45'	60.3° W
2014:10:21	00:00	09:32:32	-08°29'	61.5° W
2014:10:22	00:00	09:30:47	-09°14'	62.7° W
2014:10:23	00:00	09:28:57	-10°00'	64.0° W
2014:10:24	00:00	09:27:01	-10°48'	65.2° W
2014:10:25	00:00	09:24:58	-11°37'	66.5° W
2014:10:26	00:00	09:22:50	-12°27'	67.8° W
2014:10:27	00:00	09:20:34	-13°20'	69.0° W
2014:10:28	00:00	09:18:11	-14°13'	70.3° W
2014:10:29	00:00	09:15:41	-15°09'	71.6° W
2014:10:30	00:00	09:13:01	-16°06'	73.0° W
2014:10:31	00:00	09:10:13	-17°05'	74.3° W
2014:11:01	00:00	09:07:15	-18°06'	75.6° W
2014:11:02	00:00	09:04:07	-19°08'	77.0° W
2014:11:03	00:00	09:00:48	-20°12'	78.4° W
2014:11:04	00:00	08:57:17	-21°18'	79.7° W
2014:11:05	00:00	08:53:33	-22°26'	81.1° W
2014:11:06	00:00	08:49:35	-23°36'	82.5° W
2014:11:07	00:00	08:45:23	-24°47'	83.9° W
2014:11:08	00:00	08:40:55	-26°00'	85.3° W
2014:11:09	00:00	08:36:10	-27°14'	86.7° W
2014:11:10	00:00	08:31:07	-28°30'	88.1° W
2014:11:11	00:00	08:25:44	-29°47'	89.5° W
2014:11:12	00:00	08:20:02	-31°05'	90.8° W
2014:11:13	00:00	08:13:57	-32°24'	92.2° W
2014:11:14	00:00	08:07:29	-33°43'	93.5° W
2014:11:15	00:00	08:00:37	-35°03'	94.8° W
2014:11:16	00:00	07:53:20	-36°22'	96.1° W
2014:11:17	00:00	07:45:35	-37°41'	97.3° W
2014:11:18	00:00	07:37:23	-38°58'	98.5° W
2014:11:19	00:00	07:28:42	-40°14'	99.7° W
2014:11:20	00:00	07:19:32	-41°28'	100.8° W
2014:11:21	00:00	07:09:52	-42°39'	101.8° W
2014:11:22	00:00	06:59:44	-43°46'	102.7° W
2014:11:23	00:00	06:49:08	-44°50'	103.6° W
2014:11:24	00:00	06:38:06	-45°49'	104.4° W
2014:11:25	00:00	06:26:39	-46°44'	105.1° W
2014:11:26	00:00	06:14:50	-47°33'	105.7° W
2014:11:27	00:00	06:02:44	-48°17'	106.3° W
2014:11:28	00:00	05:50:23	-48°54'	106.7° W
2014:11:29	00:00	05:37:54	-49°26'	107.0° W
2014:11:30	00:00	05:25:21	-49°51'	107.3° W
2014:12:01	00:00	05:12:48	-50°10'	107.4° W

```
    Data         Ora         AR          Dec       Elongazione
                           [h m s]       [° ']         [ ° ]
------------------------------------------------------------------
2014:12:02      00:00     05:00:22     -50°23'      107.5°  W
2014:12:03      00:00     04:48:08     -50°29'      107.4°  W
2014:12:04      00:00     04:36:08     -50°31'      107.3°  E
2014:12:05      00:00     04:24:29     -50°27'      107.1°  E
2014:12:06      00:00     04:13:13     -50°18'      106.8°  E
2014:12:07      00:00     04:02:22     -50°05'      106.4°  E
2014:12:08      00:00     03:51:59     -49°48'      106.0°  E
2014:12:09      00:00     03:42:05     -49°28'      105.6°  E
2014:12:10      00:00     03:32:40     -49°04'      105.0°  E
2014:12:11      00:00     03:23:46     -48°38'      104.4°  E
2014:12:12      00:00     03:15:21     -48°10'      103.8°  E
2014:12:13      00:00     03:07:24     -47°41'      103.2°  E
2014:12:14      00:00     02:59:56     -47°09'      102.5°  E
2014:12:15      00:00     02:52:55     -46°37'      101.7°  E
2014:12:16      00:00     02:46:20     -46°04'      101.0°  E
2014:12:17      00:00     02:40:09     -45°30'      100.2°  E
2014:12:18      00:00     02:34:21     -44°56'       99.4°  E
2014:12:19      00:00     02:28:56     -44°22'       98.5°  E
2014:12:20      00:00     02:23:51     -43°47'       97.7°  E
2014:12:21      00:00     02:19:06     -43°13'       96.8°  E
2014:12:22      00:00     02:14:38     -42°39'       96.0°  E
2014:12:23      00:00     02:10:28     -42°05'       95.1°  E
2014:12:24      00:00     02:06:34     -41°31'       94.2°  E
2014:12:25      00:00     02:02:54     -40°58'       93.3°  E
2014:12:26      00:00     01:59:29     -40°25'       92.4°  E
2014:12:27      00:00     01:56:17     -39°53'       91.5°  E
2014:12:28      00:00     01:53:16     -39°21'       90.6°  E
2014:12:29      00:00     01:50:28     -38°50'       89.7°  E
2014:12:30      00:00     01:47:50     -38°20'       88.8°  E
2014:12:31      00:00     01:45:21     -37°50'       87.9°  E
2015:01:01      00:00     01:43:03     -37°20'       87.0°  E
2015:01:02      00:00     01:40:53     -36°51'       86.1°  E
2015:01:03      00:00     01:38:51     -36°23'       85.2°  E
2015:01:04      00:00     01:36:57     -35°55'       84.3°  E
2015:01:05      00:00     01:35:10     -35°28'       83.3°  E
2015:01:06      00:00     01:33:30     -35°02'       82.4°  E
2015:01:07      00:00     01:31:56     -34°36'       81.5°  E
2015:01:08      00:00     01:30:29     -34°11'       80.6°  E
2015:01:09      00:00     01:29:07     -33°46'       79.7°  E
2015:01:10      00:00     01:27:50     -33°21'       78.8°  E
2015:01:11      00:00     01:26:39     -32°58'       78.0°  E
2015:01:12      00:00     01:25:32     -32°34'       77.1°  E
2015:01:13      00:00     01:24:30     -32°12'       76.2°  E
```

Data	Dist(Sol) [AU]	Dist(Terra) [AU]	Magnit. [mag]
2014:07:01	1.813	2.205	8.8
2014:07:02	1.803	2.215	8.8
2014:07:03	1.793	2.224	8.8
2014:07:04	1.783	2.233	8.8
2014:07:05	1.773	2.242	8.7
2014:07:06	1.763	2.251	8.7
2014:07:07	1.753	2.260	8.7
2014:07:08	1.743	2.268	8.7
2014:07:09	1.734	2.277	8.7
2014:07:10	1.724	2.285	8.7
2014:07:11	1.714	2.293	8.6
2014:07:12	1.704	2.300	8.6
2014:07:13	1.695	2.308	8.6
2014:07:14	1.685	2.315	8.6
2014:07:15	1.676	2.322	8.6
2014:07:16	1.666	2.329	8.6
2014:07:17	1.657	2.336	8.5
2014:07:18	1.647	2.342	8.5
2014:07:19	1.638	2.348	8.5
2014:07:20	1.629	2.354	8.5
2014:07:21	1.620	2.359	8.5
2014:07:22	1.610	2.365	8.4
2014:07:23	1.601	2.370	8.4
2014:07:24	1.592	2.375	8.4
2014:07:25	1.583	2.379	8.4
2014:07:26	1.574	2.383	8.4
2014:07:27	1.566	2.387	8.3
2014:07:28	1.557	2.391	8.3
2014:07:29	1.548	2.394	8.3
2014:07:30	1.540	2.397	8.3
2014:07:31	1.531	2.400	8.3
2014:08:01	1.523	2.402	8.2
2014:08:02	1.514	2.404	8.2
2014:08:03	1.506	2.406	8.2
2014:08:04	1.498	2.407	8.2
2014:08:05	1.490	2.408	8.1
2014:08:06	1.482	2.409	8.1
2014:08:07	1.474	2.409	8.1
2014:08:08	1.466	2.409	8.1
2014:08:09	1.458	2.409	8.0
2014:08:10	1.451	2.408	8.0
2014:08:11	1.443	2.407	8.0
2014:08:12	1.436	2.406	8.0
2014:08:13	1.428	2.404	8.0
2014:08:14	1.421	2.402	7.9
2014:08:15	1.414	2.400	7.9
2014:08:16	1.407	2.397	7.9
2014:08:17	1.400	2.393	7.9
2014:08:18	1.394	2.390	7.8
2014:08:19	1.387	2.386	7.8
2014:08:20	1.380	2.382	7.8
2014:08:21	1.374	2.377	7.8

Data	Dist(Sol) [AU]	Dist(Terra) [AU]	Magnit. [mag]
2014:08:22	1.368	2.372	7.7
2014:08:23	1.362	2.366	7.7
2014:08:24	1.356	2.360	7.7
2014:08:25	1.350	2.354	7.7
2014:08:26	1.344	2.348	7.6
2014:08:27	1.338	2.341	7.6
2014:08:28	1.333	2.333	7.6
2014:08:29	1.328	2.325	7.6
2014:08:30	1.323	2.317	7.5
2014:08:31	1.318	2.309	7.5
2014:09:01	1.313	2.300	7.5
2014:09:02	1.308	2.290	7.5
2014:09:03	1.304	2.281	7.4
2014:09:04	1.299	2.271	7.4
2014:09:05	1.295	2.260	7.4
2014:09:06	1.291	2.249	7.4
2014:09:07	1.287	2.238	7.3
2014:09:08	1.283	2.226	7.3
2014:09:09	1.280	2.214	7.3
2014:09:10	1.276	2.202	7.3
2014:09:11	1.273	2.189	7.2
2014:09:12	1.270	2.176	7.2
2014:09:13	1.267	2.162	7.2
2014:09:14	1.265	2.148	7.2
2014:09:15	1.262	2.134	7.2
2014:09:16	1.260	2.119	7.1
2014:09:17	1.258	2.104	7.1
2014:09:18	1.256	2.089	7.1
2014:09:19	1.254	2.073	7.1
2014:09:20	1.252	2.057	7.0
2014:09:21	1.251	2.041	7.0
2014:09:22	1.250	2.024	7.0
2014:09:23	1.249	2.007	7.0
2014:09:24	1.248	1.990	7.0
2014:09:25	1.247	1.972	6.9
2014:09:26	1.247	1.954	6.9
2014:09:27	1.247	1.936	6.9
2014:09:28	1.247	1.917	6.9
2014:09:29	1.247	1.898	6.8
2014:09:30	1.247	1.879	6.8
2014:10:01	1.248	1.860	6.8
2014:10:02	1.248	1.840	6.8
2014:10:03	1.249	1.820	6.8
2014:10:04	1.250	1.800	6.7
2014:10:05	1.251	1.779	6.7
2014:10:06	1.253	1.759	6.7
2014:10:07	1.255	1.738	6.7
2014:10:08	1.256	1.717	6.7
2014:10:09	1.258	1.696	6.6
2014:10:10	1.261	1.674	6.6
2014:10:11	1.263	1.653	6.6
2014:10:12	1.265	1.631	6.6

Data	Dist(Sol) [AU]	Dist(Terra) [AU]	Magnit. [mag]
2014:10:13	1.268	1.609	6.6
2014:10:14	1.271	1.587	6.5
2014:10:15	1.274	1.565	6.5
2014:10:16	1.277	1.543	6.5
2014:10:17	1.281	1.521	6.5
2014:10:18	1.285	1.499	6.5
2014:10:19	1.288	1.476	6.4
2014:10:20	1.292	1.454	6.4
2014:10:21	1.296	1.432	6.4
2014:10:22	1.301	1.410	6.4
2014:10:23	1.305	1.388	6.4
2014:10:24	1.310	1.365	6.3
2014:10:25	1.314	1.344	6.3
2014:10:26	1.319	1.322	6.3
2014:10:27	1.324	1.300	6.3
2014:10:28	1.330	1.279	6.3
2014:10:29	1.335	1.258	6.3
2014:10:30	1.340	1.237	6.2
2014:10:31	1.346	1.216	6.2
2014:11:01	1.352	1.196	6.2
2014:11:02	1.358	1.176	6.2
2014:11:03	1.364	1.157	6.2
2014:11:04	1.370	1.138	6.1
2014:11:05	1.376	1.119	6.1
2014:11:06	1.383	1.102	6.1
2014:11:07	1.389	1.084	6.1
2014:11:08	1.396	1.068	6.1
2014:11:09	1.403	1.052	6.1
2014:11:10	1.409	1.037	6.1
2014:11:11	1.417	1.022	6.1
2014:11:12	1.424	1.009	6.1
2014:11:13	1.431	0.996	6.0
2014:11:14	1.438	0.985	6.0
2014:11:15	1.446	0.974	6.0
2014:11:16	1.453	0.965	6.0
2014:11:17	1.461	0.956	6.0
2014:11:18	1.469	0.949	6.1
2014:11:19	1.477	0.943	6.1
2014:11:20	1.484	0.938	6.1
2014:11:21	1.493	0.935	6.1
2014:11:22	1.501	0.933	6.1
2014:11:23	1.509	0.932	6.1
2014:11:24	1.517	0.932	6.2
2014:11:25	1.526	0.934	6.2
2014:11:26	1.534	0.937	6.2
2014:11:27	1.543	0.941	6.3
2014:11:28	1.551	0.947	6.3
2014:11:29	1.560	0.954	6.3
2014:11:30	1.569	0.962	6.4
2014:12:01	1.577	0.971	6.4
2014:12:02	1.586	0.982	6.5
2014:12:03	1.595	0.994	6.5

Data	Dist(Sol) [AU]	Dist(Terra) [AU]	Magnit. [mag]
2014:12:04	1.604	1.006	6.6
2014:12:05	1.613	1.020	6.6
2014:12:06	1.623	1.035	6.7
2014:12:07	1.632	1.051	6.7
2014:12:08	1.641	1.068	6.8
2014:12:09	1.651	1.086	6.9
2014:12:10	1.660	1.105	6.9
2014:12:11	1.669	1.125	7.0
2014:12:12	1.679	1.145	7.0
2014:12:13	1.688	1.166	7.1
2014:12:14	1.698	1.187	7.2
2014:12:15	1.708	1.210	7.2
2014:12:16	1.717	1.233	7.3
2014:12:17	1.727	1.256	7.4
2014:12:18	1.737	1.280	7.4
2014:12:19	1.747	1.304	7.5
2014:12:20	1.757	1.329	7.6
2014:12:21	1.766	1.355	7.6
2014:12:22	1.776	1.380	7.7
2014:12:23	1.786	1.406	7.8
2014:12:24	1.796	1.433	7.8
2014:12:25	1.807	1.459	7.9
2014:12:26	1.817	1.486	8.0
2014:12:27	1.827	1.514	8.0
2014:12:28	1.837	1.541	8.1
2014:12:29	1.847	1.569	8.1
2014:12:30	1.857	1.596	8.2
2014:12:31	1.868	1.624	8.3
2015:01:01	1.878	1.652	8.3
2015:01:02	1.888	1.681	8.4
2015:01:03	1.898	1.709	8.4
2015:01:04	1.909	1.737	8.5
2015:01:05	1.919	1.766	8.6
2015:01:06	1.930	1.795	8.6
2015:01:07	1.940	1.823	8.7
2015:01:08	1.951	1.852	8.7
2015:01:09	1.961	1.881	8.8
2015:01:10	1.971	1.909	8.9
2015:01:11	1.982	1.938	8.9
2015:01:12	1.993	1.967	9.0
2015:01:13	2.003	1.996	9.0

Cometa 141P/Machholz

Data	Ora	AR [h m s]	Dec [° ']	Elongazione [°]
2015:08:07	00:00	05:11:29	+37°44'	55.4° W
2015:08:08	00:00	05:19:23	+37°25'	54.8° W
2015:08:09	00:00	05:27:06	+37°04'	54.1° W
2015:08:10	00:00	05:34:38	+36°42'	53.5° W
2015:08:11	00:00	05:42:00	+36°19'	52.9° W
2015:08:12	00:00	05:49:12	+35°54'	52.3° W
2015:08:13	00:00	05:56:14	+35°29'	51.8° W
2015:08:14	00:00	06:03:06	+35°02'	51.2° W
2015:08:15	00:00	06:09:47	+34°34'	50.7° W
2015:08:16	00:00	06:16:20	+34°06'	50.3° W
2015:08:17	00:00	06:22:43	+33°37'	49.8° W
2015:08:18	00:00	06:28:56	+33°07'	49.4° W
2015:08:19	00:00	06:35:01	+32°37'	48.9° W
2015:08:20	00:00	06:40:58	+32°06'	48.5° W
2015:08:21	00:00	06:46:47	+31°35'	48.2° W
2015:08:22	00:00	06:52:27	+31°03'	47.8° W
2015:08:23	00:00	06:58:00	+30°31'	47.5° W
2015:08:24	00:00	07:03:26	+29°59'	47.2° W
2015:08:25	00:00	07:08:45	+29°26'	46.9° W
2015:08:26	00:00	07:13:57	+28°54'	46.6° W
2015:08:27	00:00	07:19:02	+28°21'	46.3° W
2015:08:28	00:00	07:24:02	+27°48'	46.1° W
2015:08:29	00:00	07:28:55	+27°16'	45.9° W
2015:08:30	00:00	07:33:43	+26°43'	45.7° W
2015:08:31	00:00	07:38:25	+26°10'	45.5° W
2015:09:01	00:00	07:43:02	+25°37'	45.3° W
2015:09:02	00:00	07:47:34	+25°04'	45.1° W
2015:09:03	00:00	07:52:01	+24°31'	45.0° W
2015:09:04	00:00	07:56:23	+23°59'	44.8° W
2015:09:05	00:00	08:00:40	+23°26'	44.7° W
2015:09:06	00:00	08:04:53	+22°54'	44.6° W
2015:09:07	00:00	08:09:02	+22°21'	44.5° W
2015:09:08	00:00	08:13:07	+21°49'	44.4° W
2015:09:09	00:00	08:17:07	+21°17'	44.4° W
2015:09:10	00:00	08:21:04	+20°45'	44.3° W

Data	Dist(Sol) [AU]	Dist(Terra) [AU]	Magnit. [mag]
2015:08:07	0.844	0.697	9.0
2015:08:08	0.837	0.705	8.9
2015:08:09	0.830	0.713	8.8
2015:08:10	0.824	0.722	8.8
2015:08:11	0.817	0.730	8.7
2015:08:12	0.811	0.739	8.6
2015:08:13	0.805	0.749	8.5
2015:08:14	0.800	0.758	8.5
2015:08:15	0.794	0.768	8.4
2015:08:16	0.790	0.778	8.4
2015:08:17	0.785	0.788	8.3
2015:08:18	0.781	0.799	8.3
2015:08:19	0.777	0.809	8.3
2015:08:20	0.773	0.820	8.2
2015:08:21	0.770	0.831	8.2
2015:08:22	0.767	0.842	8.2
2015:08:23	0.765	0.854	8.2
2015:08:24	0.763	0.865	8.2
2015:08:25	0.761	0.877	8.2
2015:08:26	0.759	0.888	8.2
2015:08:27	0.759	0.900	8.2
2015:08:28	0.758	0.912	8.2
2015:08:29	0.758	0.924	8.2
2015:08:30	0.758	0.936	8.2
2015:08:31	0.758	0.948	8.3
2015:09:01	0.759	0.960	8.3
2015:09:02	0.761	0.972	8.4
2015:09:03	0.762	0.984	8.4
2015:09:04	0.765	0.996	8.5
2015:09:05	0.767	1.008	8.6
2015:09:06	0.770	1.020	8.6
2015:09:07	0.773	1.032	8.7
2015:09:08	0.777	1.044	8.8
2015:09:09	0.780	1.056	8.9
2015:09:10	0.785	1.068	9.0

Cometa 252P/LINEAR

Data	Ora	AR [h m s]	Dec [° ']	Elongazione [°]	
2016:01:15	00:00	23:43:57	+08°00'	65.6°	E
2016:01:16	00:00	23:42:51	+07°36'	64.2°	E
2016:01:17	00:00	23:41:49	+07°14'	62.7°	E
2016:01:18	00:00	23:40:50	+06°52'	61.4°	E
2016:01:19	00:00	23:39:53	+06°31'	60.0°	E
2016:01:20	00:00	23:39:00	+06°11'	58.6°	E
2016:01:21	00:00	23:38:09	+05°51'	57.3°	E
2016:01:22	00:00	23:37:20	+05°32'	55.9°	E
2016:01:23	00:00	23:36:34	+05°13'	54.6°	E
2016:01:24	00:00	23:35:50	+04°56'	53.3°	E
2016:01:25	00:00	23:35:08	+04°38'	52.0°	E
2016:01:26	00:00	23:34:28	+04°21'	50.7°	E
2016:01:27	00:00	23:33:50	+04°05'	49.5°	E
2016:01:28	00:00	23:33:13	+03°49'	48.2°	E
2016:01:29	00:00	23:32:38	+03°34'	46.9°	E
2016:01:30	00:00	23:32:05	+03°19'	45.7°	E
2016:01:31	00:00	23:31:33	+03°04'	44.4°	E
2016:02:01	00:00	23:31:02	+02°50'	43.2°	E
2016:02:02	00:00	23:30:32	+02°36'	42.0°	E
2016:02:03	00:00	23:30:04	+02°22'	40.8°	E
2016:02:04	00:00	23:29:37	+02°09'	39.5°	E
2016:02:05	00:00	23:29:11	+01°56'	38.3°	E
2016:02:06	00:00	23:28:45	+01°44'	37.1°	E
2016:02:07	00:00	23:28:21	+01°31'	35.9°	E
2016:02:08	00:00	23:27:58	+01°19'	34.8°	E
2016:02:09	00:00	23:27:35	+01°07'	33.6°	E
2016:02:10	00:00	23:27:13	+00°55'	32.4°	E
2016:02:11	00:00	23:26:52	+00°44'	31.2°	E
2016:02:12	00:00	23:26:31	+00°33'	30.0°	E
2016:02:13	00:00	23:26:11	+00°22'	28.9°	E
2016:02:14	00:00	23:25:51	+00°11'	27.7°	E
2016:02:15	00:00	23:25:32	+00°00'	26.6°	E
2016:02:16	00:00	23:25:14	-00°11'	25.4°	E
2016:02:17	00:00	23:24:56	-00°21'	24.2°	E
2016:02:18	00:00	23:24:38	-00°32'	23.1°	E
2016:02:19	00:00	23:24:20	-00°42'	21.9°	E
2016:02:20	00:00	23:24:03	-00°52'	20.8°	E
2016:02:21	00:00	23:23:47	-01°02'	19.7°	E
2016:02:22	00:00	23:23:30	-01°12'	18.5°	E
2016:02:23	00:00	23:23:14	-01°22'	17.4°	E
2016:02:24	00:00	23:22:58	-01°31'	16.2°	E
2016:02:25	00:00	23:22:42	-01°41'	15.1°	E
2016:02:26	00:00	23:22:26	-01°51'	14.0°	E
2016:02:27	00:00	23:22:11	-02°00'	12.8°	E
2016:02:28	00:00	23:21:56	-02°10'	11.7°	E
2016:02:29	00:00	23:21:40	-02°19'	10.6°	E
2016:03:01	00:00	23:21:25	-02°28'	9.5°	E
2016:03:02	00:00	23:21:10	-02°38'	8.3°	E
2016:03:03	00:00	23:20:55	-02°47'	7.2°	E
2016:03:04	00:00	23:20:40	-02°57'	6.1°	E

85

```
Data        Ora      AR          Dec     Elongazione
                     [h m s]     [° ']    [ ° ]
-----------------------------------------------------
2016:03:05  00:00    23:20:25   -03°06'    5.0°  E
2016:03:06  00:00    23:20:10   -03°15'    3.8°  E
2016:03:07  00:00    23:19:55   -03°25'    2.7°  E
2016:03:08  00:00    23:19:40   -03°34'    1.6°  E
2016:03:09  00:00    23:19:25   -03°43'    0.7°  E
2016:03:10  00:00    23:19:10   -03°53'    0.9°  W
2016:03:11  00:00    23:18:55   -04°02'    1.9°  W
2016:03:12  00:00    23:18:40   -04°11'    3.0°  W
2016:03:13  00:00    23:18:25   -04°21'    4.1°  W
2016:03:14  00:00    23:18:10   -04°30'    5.2°  W
2016:03:15  00:00    23:17:54   -04°40'    6.3°  W
2016:03:16  00:00    23:17:39   -04°49'    7.5°  W
2016:03:17  00:00    23:17:23   -04°59'    8.6°  W
2016:03:18  00:00    23:17:08   -05°09'    9.7°  W
2016:03:19  00:00    23:16:52   -05°19'   10.8°  W
2016:03:20  00:00    23:16:36   -05°28'   12.0°  W
2016:03:21  00:00    23:16:20   -05°38'   13.1°  W
2016:03:22  00:00    23:16:03   -05°48'   14.2°  W
2016:03:23  00:00    23:15:47   -05°59'   15.3°  W
2016:03:24  00:00    23:15:30   -06°09'   16.5°  W
2016:03:25  00:00    23:15:14   -06°19'   17.6°  W
2016:03:26  00:00    23:14:56   -06°30'   18.7°  W
2016:03:27  00:00    23:14:39   -06°40'   19.9°  W
2016:03:28  00:00    23:14:22   -06°51'   21.0°  W
2016:03:29  00:00    23:14:04   -07°02'   22.1°  W
2016:03:30  00:00    23:13:46   -07°13'   23.3°  W
2016:03:31  00:00    23:13:27   -07°24'   24.4°  W
2016:04:01  00:00    23:13:09   -07°35'   25.5°  W
2016:04:02  00:00    23:12:50   -07°47'   26.7°  W
2016:04:03  00:00    23:12:30   -07°59'   27.8°  W
2016:04:04  00:00    23:12:10   -08°11'   29.0°  W
2016:04:05  00:00    23:11:50   -08°23'   30.1°  W
2016:04:06  00:00    23:11:29   -08°36'   31.3°  W
2016:04:07  00:00    23:11:08   -08°48'   32.4°  W
2016:04:08  00:00    23:10:46   -09°01'   33.6°  W
2016:04:09  00:00    23:10:24   -09°15'   34.7°  W
2016:04:10  00:00    23:10:01   -09°28'   35.9°  W
2016:04:11  00:00    23:09:37   -09°42'   37.1°  W
2016:04:12  00:00    23:09:13   -09°56'   38.3°  W
2016:04:13  00:00    23:08:48   -10°11'   39.4°  W
2016:04:14  00:00    23:08:21   -10°26'   40.6°  W
2016:04:15  00:00    23:07:54   -10°41'   41.8°  W
2016:04:16  00:00    23:07:26   -10°57'   43.0°  W
2016:04:17  00:00    23:06:57   -11°14'   44.2°  W
2016:04:18  00:00    23:06:27   -11°31'   45.4°  W
2016:04:19  00:00    23:05:55   -11°48'   46.6°  W
2016:04:20  00:00    23:05:22   -12°06'   47.9°  W
2016:04:21  00:00    23:04:48   -12°25'   49.1°  W
2016:04:22  00:00    23:04:12   -12°44'   50.4°  W
2016:04:23  00:00    23:03:34   -13°04'   51.6°  W
2016:04:24  00:00    23:02:54   -13°25'   52.9°  W
2016:04:25  00:00    23:02:12   -13°46'   54.2°  W
```

```
   Data        Ora       AR       Dec   Elongazione
                        [h m s]    [° ']    [ ° ]
-------------------------------------------------
2016:04:26    00:00   23:01:28  -14°09'   55.5° W
2016:04:27    00:00   23:00:42  -14°32'   56.8° W
2016:04:28    00:00   22:59:53  -14°56'   58.1° W
2016:04:29    00:00   22:59:01  -15°22'   59.4° W
2016:04:30    00:00   22:58:05  -15°48'   60.8° W
2016:05:01    00:00   22:57:07  -16°16'   62.2° W
2016:05:02    00:00   22:56:04  -16°45'   63.6° W
2016:05:03    00:00   22:54:58  -17°15'   65.0° W
2016:05:04    00:00   22:53:47  -17°47'   66.4° W
2016:05:05    00:00   22:52:30  -18°21'   67.9° W
2016:05:06    00:00   22:51:09  -18°57'   69.4° W
2016:05:07    00:00   22:49:41  -19°34'   71.0° W
2016:05:08    00:00   22:48:07  -20°13'   72.5° W
2016:05:09    00:00   22:46:25  -20°55'   74.1° W
2016:05:10    00:00   22:44:35  -21°39'   75.8° W
2016:05:11    00:00   22:42:36  -22°26'   77.5° W
2016:05:12    00:00   22:40:26  -23°15'   79.2° W
2016:05:13    00:00   22:38:05  -24°08'   81.0° W
2016:05:14    00:00   22:35:31  -25°04'   82.9° W
2016:05:15    00:00   22:32:43  -26°03'   84.8° W
2016:05:16    00:00   22:29:38  -27°07'   86.8° W
2016:05:17    00:00   22:26:14  -28°14'   88.8° W
2016:05:18    00:00   22:22:29  -29°26'   91.0° W
2016:05:19    00:00   22:18:20  -30°43'   93.2° W
2016:05:20    00:00   22:13:43  -32°05'   95.5° W
2016:05:21    00:00   22:08:33  -33°32'   97.9° W
2016:05:22    00:00   22:02:46  -35°05'  100.4° W
2016:05:23    00:00   21:56:15  -36°43'  103.1° W
2016:05:24    00:00   21:48:53  -38°27'  105.8° W
2016:05:25    00:00   21:40:32  -40°17'  108.7° W
2016:05:26    00:00   21:30:59  -42°12'  111.6° W
2016:05:27    00:00   21:20:04  -44°11'  114.7° W
2016:05:28    00:00   21:07:31  -46°14'  117.9° W
2016:05:29    00:00   20:53:04  -48°18'  121.2° W
2016:05:30    00:00   20:36:26  -50°21'  124.5° W
2016:05:31    00:00   20:17:19  -52°19'  127.8° W
2016:06:01    00:00   19:55:30  -54°08'  131.1° W
2016:06:02    00:00   19:30:53  -55°44'  134.3° W
2016:06:03    00:00   19:03:37  -57°01'  137.2° W
2016:06:04    00:00   18:34:11  -57°55'  139.9° W
2016:06:05    00:00   18:03:27  -58°21'  142.1° W
2016:06:06    00:00   17:32:30  -58°18'  143.8° W
2016:06:07    00:00   17:02:31  -57°48'  145.0° W
2016:06:08    00:00   16:34:25  -56°53'  145.5° E
2016:06:09    00:00   16:08:51  -55°38'  145.4° E
2016:06:10    00:00   15:46:04  -54°08'  144.7° E
2016:06:11    00:00   15:26:04  -52°28'  143.6° E
2016:06:12    00:00   15:08:38  -50°42'  142.1° E
2016:06:13    00:00   14:53:31  -48°54'  140.4° E
2016:06:14    00:00   14:40:25  -47°07'  138.5° E
2016:06:15    00:00   14:29:02  -45°21'  136.5° E
2016:06:16    00:00   14:19:08  -43°39'  134.4° E
```

Data	Ora	AR [h m s]	Dec [° ']	Elongazione [°]
2016:06:17	00:00	14:10:29	-42°02'	132.3° E
2016:06:18	00:00	14:02:55	-40°29'	130.2° E
2016:06:19	00:00	13:56:15	-39°01'	128.2° E
2016:06:20	00:00	13:50:22	-37°38'	126.2° E

Data	Dist(Sol) [AU]	Dist(Terra) [AU]	Magnit. [mag]
2016:01:15	1.350	1.416	9.0
2016:01:16	1.341	1.436	9.0
2016:01:17	1.333	1.456	8.9
2016:01:18	1.325	1.476	8.9
2016:01:19	1.317	1.496	8.9
2016:01:20	1.308	1.515	8.8
2016:01:21	1.300	1.535	8.8
2016:01:22	1.292	1.554	8.7
2016:01:23	1.284	1.573	8.7
2016:01:24	1.276	1.591	8.7
2016:01:25	1.268	1.609	8.6
2016:01:26	1.261	1.627	8.6
2016:01:27	1.253	1.645	8.5
2016:01:28	1.245	1.662	8.5
2016:01:29	1.237	1.679	8.4
2016:01:30	1.230	1.696	8.4
2016:01:31	1.222	1.713	8.3
2016:02:01	1.215	1.729	8.3
2016:02:02	1.208	1.744	8.3
2016:02:03	1.200	1.760	8.2
2016:02:04	1.193	1.775	8.2
2016:02:05	1.186	1.789	8.1
2016:02:06	1.179	1.804	8.1
2016:02:07	1.172	1.817	8.0
2016:02:08	1.165	1.831	8.0
2016:02:09	1.159	1.844	7.9

Data	Dist(Sol) [AU]	Dist(Terra) [AU]	Magnit. [mag]
2016:02:10	1.152	1.857	7.9
2016:02:11	1.145	1.869	7.8
2016:02:12	1.139	1.880	7.8
2016:02:13	1.133	1.892	7.7
2016:02:14	1.126	1.903	7.7
2016:02:15	1.120	1.913	7.6
2016:02:16	1.114	1.923	7.6
2016:02:17	1.108	1.932	7.5
2016:02:18	1.103	1.941	7.5
2016:02:19	1.097	1.950	7.5
2016:02:20	1.092	1.958	7.4
2016:02:21	1.086	1.965	7.4
2016:02:22	1.081	1.972	7.3
2016:02:23	1.076	1.979	7.3
2016:02:24	1.071	1.984	7.2
2016:02:25	1.066	1.990	7.2
2016:02:26	1.061	1.995	7.1
2016:02:27	1.057	1.999	7.1
2016:02:28	1.053	2.003	7.1
2016:02:29	1.048	2.006	7.0
2016:03:01	1.044	2.009	7.0
2016:03:02	1.041	2.011	6.9
2016:03:03	1.037	2.013	6.9
2016:03:04	1.033	2.014	6.9
2016:03:05	1.030	2.015	6.8
2016:03:06	1.027	2.015	6.8
2016:03:07	1.024	2.014	6.8
2016:03:08	1.021	2.013	6.7
2016:03:09	1.018	2.011	6.7
2016:03:10	1.016	2.009	6.7
2016:03:11	1.014	2.006	6.7
2016:03:12	1.012	2.003	6.6
2016:03:13	1.010	1.999	6.6
2016:03:14	1.008	1.994	6.6
2016:03:15	1.006	1.989	6.6
2016:03:16	1.005	1.983	6.5
2016:03:17	1.004	1.977	6.5
2016:03:18	1.003	1.970	6.5
2016:03:19	1.002	1.963	6.5
2016:03:20	1.002	1.955	6.5
2016:03:21	1.002	1.946	6.5
2016:03:22	1.001	1.937	6.5
2016:03:23	1.001	1.927	6.4
2016:03:24	1.002	1.917	6.4
2016:03:25	1.002	1.906	6.4
2016:03:26	1.003	1.895	6.4
2016:03:27	1.004	1.883	6.4
2016:03:28	1.005	1.871	6.4
2016:03:29	1.006	1.858	6.4
2016:03:30	1.007	1.845	6.4
2016:03:31	1.009	1.831	6.4
2016:04:01	1.011	1.816	6.4

Data	Dist(Sol) [AU]	Dist(Terra) [AU]	Magnit. [mag]
2016:04:02	1.013	1.801	6.4
2016:04:03	1.015	1.786	6.4
2016:04:04	1.018	1.770	6.4
2016:04:05	1.020	1.753	6.4
2016:04:06	1.023	1.736	6.4
2016:04:07	1.026	1.719	6.5
2016:04:08	1.029	1.701	6.5
2016:04:09	1.032	1.683	6.5
2016:04:10	1.036	1.664	6.5
2016:04:11	1.039	1.645	6.5
2016:04:12	1.043	1.625	6.5
2016:04:13	1.047	1.605	6.5
2016:04:14	1.051	1.585	6.5
2016:04:15	1.056	1.564	6.6
2016:04:16	1.060	1.543	6.6
2016:04:17	1.065	1.521	6.6
2016:04:18	1.069	1.500	6.6
2016:04:19	1.074	1.477	6.6
2016:04:20	1.079	1.455	6.6
2016:04:21	1.084	1.432	6.7
2016:04:22	1.090	1.408	6.7
2016:04:23	1.095	1.385	6.7
2016:04:24	1.101	1.361	6.7
2016:04:25	1.106	1.337	6.7
2016:04:26	1.112	1.312	6.7
2016:04:27	1.118	1.288	6.8
2016:04:28	1.124	1.263	6.8
2016:04:29	1.131	1.237	6.8
2016:04:30	1.137	1.212	6.8
2016:05:01	1.143	1.187	6.8
2016:05:02	1.150	1.161	6.8
2016:05:03	1.156	1.135	6.9
2016:05:04	1.163	1.109	6.9
2016:05:05	1.170	1.083	6.9
2016:05:06	1.177	1.057	6.9
2016:05:07	1.184	1.030	6.9
2016:05:08	1.191	1.004	6.9
2016:05:09	1.198	0.977	6.9
2016:05:10	1.205	0.951	6.9
2016:05:11	1.213	0.925	6.9
2016:05:12	1.220	0.898	6.9
2016:05:13	1.227	0.872	6.9
2016:05:14	1.235	0.846	6.9
2016:05:15	1.243	0.820	6.9
2016:05:16	1.250	0.795	6.9
2016:05:17	1.258	0.769	6.9
2016:05:18	1.266	0.744	6.9
2016:05:19	1.274	0.720	6.9
2016:05:20	1.282	0.696	6.9
2016:05:21	1.290	0.672	6.9
2016:05:22	1.298	0.649	6.9
2016:05:23	1.306	0.627	6.9

Data	Dist(Sol) [AU]	Dist(Terra) [AU]	Magnit. [mag]
2016:05:24	1.314	0.606	6.9
2016:05:25	1.322	0.585	6.9
2016:05:26	1.330	0.566	6.9
2016:05:27	1.339	0.548	6.9
2016:05:28	1.347	0.532	6.9
2016:05:29	1.355	0.517	6.9
2016:05:30	1.364	0.504	6.9
2016:05:31	1.372	0.492	6.9
2016:06:01	1.380	0.483	6.9
2016:06:02	1.389	0.476	7.0
2016:06:03	1.398	0.471	7.0
2016:06:04	1.406	0.469	7.1
2016:06:05	1.415	0.469	7.1
2016:06:06	1.423	0.472	7.2
2016:06:07	1.432	0.477	7.3
2016:06:08	1.441	0.484	7.4
2016:06:09	1.449	0.494	7.5
2016:06:10	1.458	0.506	7.6
2016:06:11	1.467	0.520	7.7
2016:06:12	1.475	0.536	7.9
2016:06:13	1.484	0.553	8.0
2016:06:14	1.493	0.572	8.1
2016:06:15	1.502	0.592	8.3
2016:06:16	1.511	0.614	8.4
2016:06:17	1.519	0.637	8.6
2016:06:18	1.528	0.660	8.7
2016:06:19	1.537	0.685	8.8
2016:06:20	1.546	0.711	9.0

Cometa 2P/Encke

Data	Ora	AR [h m s]	Dec [° ']	Elongazione [°]	
2017:02:15	00:00	23:58:05	+07°34'	36.7°	E
2017:02:16	00:00	23:59:27	+07°36'	36.0°	E
2017:02:17	00:00	00:00:46	+07°37'	35.3°	E
2017:02:18	00:00	00:02:01	+07°37'	34.6°	E
2017:02:19	00:00	00:03:12	+07°35'	33.8°	E
2017:02:20	00:00	00:04:18	+07°32'	33.0°	E
2017:02:21	00:00	00:05:17	+07°28'	32.2°	E
2017:02:22	00:00	00:06:09	+07°22'	31.3°	E
2017:02:23	00:00	00:06:52	+07°13'	30.4°	E
2017:02:24	00:00	00:07:24	+07°03'	29.4°	E
2017:02:25	00:00	00:07:43	+06°49'	28.4°	E
2017:02:26	00:00	00:07:48	+06°32'	27.3°	E
2017:02:27	00:00	00:07:34	+06°11'	26.0°	E
2017:02:28	00:00	00:07:01	+05°46'	24.7°	E
2017:03:01	00:00	00:06:04	+05°16'	23.3°	E
2017:03:02	00:00	00:04:40	+04°40'	21.7°	E
2017:03:03	00:00	00:02:47	+03°59'	19.9°	E
2017:03:04	00:00	00:00:21	+03°10'	18.0°	E
2017:03:05	00:00	23:57:19	+02°15'	15.8°	E
2017:03:06	00:00	23:53:42	+01°13'	13.5°	E
2017:03:07	00:00	23:49:28	+00°04'	11.1°	E
2017:03:08	00:00	23:44:41	-01°11'	8.4°	E
2017:03:09	00:00	23:39:24	-02°31'	5.7°	E
2017:03:10	00:00	23:33:44	-03°54'	3.0°	E
2017:03:11	00:00	23:27:50	-05°19'	1.7°	E
2017:03:12	00:00	23:21:51	-06°42'	3.8°	W
2017:03:13	00:00	23:15:55	-08°03'	6.6°	W
2017:03:14	00:00	23:10:11	-09°20'	9.4°	W
2017:03:15	00:00	23:04:47	-10°31'	12.1°	W
2017:03:16	00:00	22:59:46	-11°36'	14.7°	W
2017:03:17	00:00	22:55:12	-12°35'	17.1°	W
2017:03:18	00:00	22:51:07	-13°27'	19.4°	W
2017:03:19	00:00	22:47:31	-14°12'	21.5°	W
2017:03:20	00:00	22:44:22	-14°52'	23.5°	W
2017:03:21	00:00	22:41:40	-15°26'	25.3°	W
2017:03:22	00:00	22:39:22	-15°56'	27.0°	W
2017:03:23	00:00	22:37:27	-16°21'	28.6°	W
2017:03:24	00:00	22:35:52	-16°43'	30.0°	W
2017:03:25	00:00	22:34:35	-17°01'	31.4°	W
2017:03:26	00:00	22:33:34	-17°16'	32.7°	W
2017:03:27	00:00	22:32:47	-17°29'	34.0°	W
2017:03:28	00:00	22:32:13	-17°39'	35.1°	W
2017:03:29	00:00	22:31:50	-17°48'	36.2°	W
2017:03:30	00:00	22:31:37	-17°55'	37.3°	W
2017:03:31	00:00	22:31:32	-18°00'	38.3°	W
2017:04:01	00:00	22:31:35	-18°04'	39.3°	W
2017:04:02	00:00	22:31:44	-18°08'	40.3°	W
2017:04:03	00:00	22:31:58	-18°10'	41.2°	W
2017:04:04	00:00	22:32:17	-18°11'	42.1°	W

Data	Dist(Sol) [AU]	Dist(Terra) [AU]	Magnit. [mag]
2017:02:15	0.653	1.071	8.9
2017:02:16	0.634	1.054	8.6
2017:02:17	0.616	1.037	8.4
2017:02:18	0.597	1.020	8.2
2017:02:19	0.579	1.001	7.9
2017:02:20	0.560	0.983	7.7
2017:02:21	0.542	0.964	7.4
2017:02:22	0.524	0.944	7.2
2017:02:23	0.506	0.924	6.9
2017:02:24	0.488	0.904	6.6
2017:02:25	0.471	0.883	6.3
2017:02:26	0.454	0.862	6.0
2017:02:27	0.438	0.841	5.7
2017:02:28	0.422	0.819	5.4
2017:03:01	0.407	0.798	5.2
2017:03:02	0.393	0.777	4.9
2017:03:03	0.380	0.757	4.6
2017:03:04	0.368	0.738	4.3
2017:03:05	0.358	0.719	4.1
2017:03:06	0.350	0.703	3.9
2017:03:07	0.343	0.688	3.7
2017:03:08	0.339	0.676	3.6
2017:03:09	0.336	0.666	3.5
2017:03:10	0.336	0.659	3.5
2017:03:11	0.338	0.656	3.5
2017:03:12	0.343	0.655	3.6
2017:03:13	0.349	0.658	3.7
2017:03:14	0.357	0.663	3.9
2017:03:15	0.367	0.671	4.1
2017:03:16	0.378	0.681	4.3
2017:03:17	0.391	0.692	4.6
2017:03:18	0.405	0.705	4.9
2017:03:19	0.420	0.719	5.1
2017:03:20	0.435	0.734	5.4
2017:03:21	0.452	0.750	5.7
2017:03:22	0.468	0.766	6.0
2017:03:23	0.486	0.782	6.3
2017:03:24	0.503	0.798	6.5
2017:03:25	0.521	0.814	6.8
2017:03:26	0.539	0.830	7.1
2017:03:27	0.558	0.845	7.3
2017:03:28	0.576	0.861	7.6
2017:03:29	0.595	0.876	7.8
2017:03:30	0.613	0.891	8.1
2017:03:31	0.632	0.905	8.3
2017:04:01	0.650	0.919	8.5
2017:04:02	0.669	0.933	8.7
2017:04:03	0.687	0.946	8.9
2017:04:04	0.705	0.959	9.1

46P/Wirtanen

Data	Ora	AR [h m s]	Dec [° ']	Elongazione [°]	
2018:10:10	00:00	01:51:47	-29°24'	142.1°	W
2018:10:11	00:00	01:51:55	-29°41'	141.7°	W
2018:10:12	00:00	01:52:01	-29°57'	141.3°	W
2018:10:13	00:00	01:52:06	-30°13'	140.9°	W
2018:10:14	00:00	01:52:10	-30°29'	140.5°	W
2018:10:15	00:00	01:52:13	-30°45'	140.1°	W
2018:10:16	00:00	01:52:15	-31°00'	139.6°	W
2018:10:17	00:00	01:52:16	-31°14'	139.2°	W
2018:10:18	00:00	01:52:17	-31°28'	138.7°	W
2018:10:19	00:00	01:52:17	-31°42'	138.2°	W
2018:10:20	00:00	01:52:16	-31°55'	137.7°	W
2018:10:21	00:00	01:52:15	-32°07'	137.2°	W
2018:10:22	00:00	01:52:14	-32°19'	136.7°	W
2018:10:23	00:00	01:52:12	-32°30'	136.2°	W
2018:10:24	00:00	01:52:10	-32°41'	135.7°	E
2018:10:25	00:00	01:52:09	-32°50'	135.1°	E
2018:10:26	00:00	01:52:07	-32°59'	134.6°	E
2018:10:27	00:00	01:52:05	-33°08'	134.1°	E
2018:10:28	00:00	01:52:04	-33°15'	133.5°	E
2018:10:29	00:00	01:52:04	-33°21'	133.0°	E
2018:10:30	00:00	01:52:04	-33°27'	132.5°	E
2018:10:31	00:00	01:52:05	-33°32'	132.0°	E
2018:11:01	00:00	01:52:07	-33°35'	131.5°	E
2018:11:02	00:00	01:52:09	-33°38'	130.9°	E
2018:11:03	00:00	01:52:14	-33°39'	130.4°	E
2018:11:04	00:00	01:52:20	-33°39'	129.9°	E
2018:11:05	00:00	01:52:27	-33°39'	129.5°	E
2018:11:06	00:00	01:52:36	-33°36'	129.0°	E
2018:11:07	00:00	01:52:48	-33°33'	128.5°	E
2018:11:08	00:00	01:53:01	-33°28'	128.1°	E
2018:11:09	00:00	01:53:18	-33°22'	127.6°	E
2018:11:10	00:00	01:53:36	-33°14'	127.2°	E
2018:11:11	00:00	01:53:58	-33°05'	126.8°	E
2018:11:12	00:00	01:54:23	-32°54'	126.5°	E
2018:11:13	00:00	01:54:52	-32°41'	126.1°	E
2018:11:14	00:00	01:55:24	-32°27'	125.8°	E
2018:11:15	00:00	01:56:00	-32°10'	125.5°	E
2018:11:16	00:00	01:56:40	-31°51'	125.2°	E
2018:11:17	00:00	01:57:25	-31°31'	125.0°	E
2018:11:18	00:00	01:58:15	-31°07'	124.8°	E
2018:11:19	00:00	01:59:10	-30°42'	124.6°	E
2018:11:20	00:00	02:00:11	-30°13'	124.5°	E
2018:11:21	00:00	02:01:18	-29°42'	124.4°	E
2018:11:22	00:00	02:02:31	-29°08'	124.4°	E
2018:11:23	00:00	02:03:50	-28°30'	124.4°	E
2018:11:24	00:00	02:05:17	-27°49'	124.5°	E
2018:11:25	00:00	02:06:52	-27°04'	124.7°	E
2018:11:26	00:00	02:08:35	-26°14'	124.9°	E
2018:11:27	00:00	02:10:27	-25°20'	125.1°	E
2018:11:28	00:00	02:12:28	-24°21'	125.5°	E

Data	Ora	AR [h m s]	Dec [° ']	Elongazione [°]
2018:11:29	00:00	02:14:40	-23°17'	125.9° E
2018:11:30	00:00	02:17:02	-22°06'	126.5° E
2018:12:01	00:00	02:19:37	-20°49'	127.1° E
2018:12:02	00:00	02:22:23	-19°25'	127.9° E
2018:12:03	00:00	02:25:23	-17°53'	128.7° E
2018:12:04	00:00	02:28:38	-16°13'	129.7° E
2018:12:05	00:00	02:32:08	-14°23'	130.9° E
2018:12:06	00:00	02:35:55	-12°24'	132.1° E
2018:12:07	00:00	02:40:01	-10°14'	133.6° E
2018:12:08	00:00	02:44:25	-07°52'	135.2° E
2018:12:09	00:00	02:49:11	-05°19'	136.9° E
2018:12:10	00:00	02:54:19	-02°33'	138.8° E
2018:12:11	00:00	02:59:51	+00°25'	140.8° E
2018:12:12	00:00	03:05:50	+03°35'	143.0° E
2018:12:13	00:00	03:12:16	+06°57'	145.2° E
2018:12:14	00:00	03:19:11	+10°29'	147.4° E
2018:12:15	00:00	03:26:38	+14°09'	149.6° E
2018:12:16	00:00	03:34:38	+17°54'	151.7° E
2018:12:17	00:00	03:43:12	+21°43'	153.4° E
2018:12:18	00:00	03:52:21	+25°31'	154.9° E
2018:12:19	00:00	04:02:07	+29°15'	155.8° E
2018:12:20	00:00	04:12:29	+32°52'	156.3° E
2018:12:21	00:00	04:23:27	+36°19'	156.3° E
2018:12:22	00:00	04:34:59	+39°32'	155.8° E
2018:12:23	00:00	04:47:03	+42°32'	155.0° E
2018:12:24	00:00	04:59:35	+45°15'	154.0° E
2018:12:25	00:00	05:12:30	+47°43'	152.8° E
2018:12:26	00:00	05:25:44	+49°54'	151.6° E
2018:12:27	00:00	05:39:09	+51°49'	150.3° E
2018:12:28	00:00	05:52:38	+53°29'	149.1° E
2018:12:29	00:00	06:06:04	+54°54'	148.0° E
2018:12:30	00:00	06:19:20	+56°07'	146.9° E
2018:12:31	00:00	06:32:19	+57°08'	146.0° E
2019:01:01	00:00	06:44:55	+57°58'	145.1° W
2019:01:02	00:00	06:57:03	+58°39'	144.3° W
2019:01:03	00:00	07:08:38	+59°11'	143.6° W
2019:01:04	00:00	07:19:38	+59°37'	143.0° W
2019:01:05	00:00	07:30:02	+59°56'	142.4° W
2019:01:06	00:00	07:39:47	+60°09'	142.0° W
2019:01:07	00:00	07:48:54	+60°18'	141.6° W
2019:01:08	00:00	07:57:24	+60°23'	141.2° W
2019:01:09	00:00	08:05:18	+60°25'	140.9° W
2019:01:10	00:00	08:12:37	+60°24'	140.7° W
2019:01:11	00:00	08:19:22	+60°20'	140.5° W
2019:01:12	00:00	08:25:36	+60°15'	140.4° W
2019:01:13	00:00	08:31:21	+60°07'	140.3° W
2019:01:14	00:00	08:36:38	+59°58'	140.2° W
2019:01:15	00:00	08:41:30	+59°48'	140.2° W
2019:01:16	00:00	08:45:57	+59°37'	140.2° W
2019:01:17	00:00	08:50:03	+59°24'	140.2° W
2019:01:18	00:00	08:53:48	+59°11'	140.2° W
2019:01:19	00:00	08:57:14	+58°58'	140.3° W

```
   Data        Ora        AR         Dec      Elongazione
                        [h m s]      [° ']       [ ° ]
-------------------------------------------------------
2019:01:20    00:00    09:00:24    +58°43'    140.3°  W
2019:01:21    00:00    09:03:17    +58°28'    140.4°  W
2019:01:22    00:00    09:05:56    +58°13'    140.5°  W
2019:01:23    00:00    09:08:21    +57°57'    140.6°  W
2019:01:24    00:00    09:10:34    +57°40'    140.7°  W
2019:01:25    00:00    09:12:36    +57°24'    140.8°  W
2019:01:26    00:00    09:14:28    +57°07'    140.9°  W
2019:01:27    00:00    09:16:10    +56°50'    141.0°  W
2019:01:28    00:00    09:17:43    +56°32'    141.2°  W
2019:01:29    00:00    09:19:09    +56°14'    141.3°  W
2019:01:30    00:00    09:20:28    +55°56'    141.4°  W
2019:01:31    00:00    09:21:40    +55°38'    141.5°  W
2019:02:01    00:00    09:22:47    +55°19'    141.6°  W
2019:02:02    00:00    09:23:48    +55°01'    141.7°  W
2019:02:03    00:00    09:24:44    +54°42'    141.8°  W
2019:02:04    00:00    09:25:37    +54°23'    141.9°  W
2019:02:05    00:00    09:26:25    +54°04'    141.9°  W
2019:02:06    00:00    09:27:10    +53°44'    142.0°  W
2019:02:07    00:00    09:27:52    +53°25'    142.0°  W
2019:02:08    00:00    09:28:31    +53°05'    142.1°  W
2019:02:09    00:00    09:29:07    +52°45'    142.1°  E
2019:02:10    00:00    09:29:42    +52°26'    142.1°  E
2019:02:11    00:00    09:30:15    +52°06'    142.1°  E
2019:02:12    00:00    09:30:46    +51°45'    142.0°  E
2019:02:13    00:00    09:31:16    +51°25'    142.0°  E
2019:02:14    00:00    09:31:45    +51°05'    141.9°  E
```

Data	Dist(Sol) [AU]	Dist(Terra) [AU]	Magnit. [mag]
2018:10:10	1.349	0.413	9.0
2018:10:11	1.341	0.406	9.0
2018:10:12	1.333	0.399	8.9
2018:10:13	1.326	0.392	8.8
2018:10:14	1.318	0.386	8.7
2018:10:15	1.311	0.379	8.7
2018:10:16	1.304	0.373	8.6
2018:10:17	1.296	0.366	8.5
2018:10:18	1.289	0.360	8.4
2018:10:19	1.282	0.354	8.4
2018:10:20	1.275	0.348	8.3
2018:10:21	1.268	0.341	8.2
2018:10:22	1.261	0.335	8.1
2018:10:23	1.254	0.329	8.1
2018:10:24	1.247	0.323	8.0
2018:10:25	1.240	0.318	7.9
2018:10:26	1.233	0.312	7.8
2018:10:27	1.227	0.306	7.8
2018:10:28	1.220	0.300	7.7
2018:10:29	1.214	0.295	7.6
2018:10:30	1.207	0.289	7.5
2018:10:31	1.201	0.284	7.5
2018:11:01	1.195	0.278	7.4
2018:11:02	1.189	0.272	7.3
2018:11:03	1.183	0.267	7.2
2018:11:04	1.177	0.262	7.2
2018:11:05	1.171	0.256	7.1
2018:11:06	1.166	0.251	7.0
2018:11:07	1.160	0.245	6.9
2018:11:08	1.155	0.240	6.8
2018:11:09	1.149	0.235	6.8
2018:11:10	1.144	0.230	6.7
2018:11:11	1.139	0.224	6.6
2018:11:12	1.134	0.219	6.5
2018:11:13	1.129	0.214	6.4
2018:11:14	1.124	0.209	6.4
2018:11:15	1.120	0.204	6.3
2018:11:16	1.115	0.198	6.2
2018:11:17	1.111	0.193	6.1
2018:11:18	1.107	0.188	6.0
2018:11:19	1.103	0.183	5.9
2018:11:20	1.099	0.178	5.9
2018:11:21	1.095	0.173	5.8
2018:11:22	1.091	0.168	5.7
2018:11:23	1.088	0.163	5.6
2018:11:24	1.085	0.158	5.5
2018:11:25	1.082	0.153	5.4
2018:11:26	1.079	0.148	5.3
2018:11:27	1.076	0.143	5.3
2018:11:28	1.073	0.138	5.2
2018:11:29	1.070	0.134	5.1
2018:11:30	1.068	0.129	5.0

Data	Dist(Sol) [AU]	Dist(Terra) [AU]	Magnit. [mag]
2018:12:01	1.066	0.124	4.9
2018:12:02	1.064	0.120	4.8
2018:12:03	1.062	0.116	4.7
2018:12:04	1.060	0.111	4.6
2018:12:05	1.059	0.107	4.5
2018:12:06	1.057	0.103	4.4
2018:12:07	1.056	0.099	4.3
2018:12:08	1.055	0.096	4.3
2018:12:09	1.054	0.092	4.2
2018:12:10	1.054	0.089	4.1
2018:12:11	1.053	0.086	4.0
2018:12:12	1.053	0.084	4.0
2018:12:13	1.053	0.082	3.9
2018:12:14	1.053	0.080	3.9
2018:12:15	1.053	0.079	3.8
2018:12:16	1.053	0.078	3.8
2018:12:17	1.054	0.077	3.8
2018:12:18	1.055	0.078	3.8
2018:12:19	1.056	0.078	3.8
2018:12:20	1.057	0.079	3.8
2018:12:21	1.058	0.081	3.9
2018:12:22	1.059	0.082	4.0
2018:12:23	1.061	0.085	4.0
2018:12:24	1.063	0.087	4.1
2018:12:25	1.065	0.090	4.2
2018:12:26	1.067	0.094	4.3
2018:12:27	1.069	0.097	4.4
2018:12:28	1.072	0.101	4.5
2018:12:29	1.074	0.105	4.6
2018:12:30	1.077	0.110	4.7
2018:12:31	1.080	0.114	4.8
2019:01:01	1.083	0.119	4.9
2019:01:02	1.086	0.124	5.0
2019:01:03	1.090	0.129	5.1
2019:01:04	1.093	0.134	5.2
2019:01:05	1.097	0.139	5.3
2019:01:06	1.101	0.144	5.4
2019:01:07	1.105	0.150	5.5
2019:01:08	1.109	0.155	5.6
2019:01:09	1.113	0.161	5.7
2019:01:10	1.117	0.167	5.8
2019:01:11	1.122	0.172	5.9
2019:01:12	1.127	0.178	6.0
2019:01:13	1.131	0.184	6.1
2019:01:14	1.136	0.190	6.2
2019:01:15	1.141	0.196	6.3
2019:01:16	1.146	0.202	6.4
2019:01:17	1.152	0.209	6.5
2019:01:18	1.157	0.215	6.6
2019:01:19	1.163	0.221	6.7
2019:01:20	1.168	0.228	6.8
2019:01:21	1.174	0.234	6.9

Data	Dist(Sol) [AU]	Dist(Terra) [AU]	Magnit. [mag]
2019:01:22	1.180	0.241	7.0
2019:01:23	1.186	0.247	7.1
2019:01:24	1.192	0.254	7.2
2019:01:25	1.198	0.261	7.3
2019:01:26	1.204	0.267	7.3
2019:01:27	1.210	0.274	7.4
2019:01:28	1.217	0.281	7.5
2019:01:29	1.223	0.288	7.6
2019:01:30	1.230	0.295	7.7
2019:01:31	1.236	0.303	7.8
2019:02:01	1.243	0.310	7.9
2019:02:02	1.250	0.317	8.0
2019:02:03	1.257	0.325	8.0
2019:02:04	1.264	0.332	8.1
2019:02:05	1.271	0.340	8.2
2019:02:06	1.278	0.348	8.3
2019:02:07	1.285	0.355	8.4
2019:02:08	1.292	0.363	8.5
2019:02:09	1.300	0.371	8.6
2019:02:10	1.307	0.379	8.6
2019:02:11	1.314	0.388	8.7
2019:02:12	1.322	0.396	8.8
2019:02:13	1.329	0.404	8.9
2019:02:14	1.337	0.413	9.0

Cometa 2P/Encke

Data	Ora	AR [h m s]	Dec [° ']	Elongazione [°]
2020:06:07	00:00	05:00:42	+26°56'	4.2° W
2020:06:08	00:00	05:07:22	+26°57'	4.1° E
2020:06:09	00:00	05:14:10	+26°56'	4.1° E
2020:06:10	00:00	05:21:09	+26°55'	4.1° E
2020:06:11	00:00	05:28:16	+26°51'	4.3° E
2020:06:12	00:00	05:35:33	+26°46'	4.6° E
2020:06:13	00:00	05:43:00	+26°38'	5.0° E
2020:06:14	00:00	05:50:35	+26°29'	5.4° E
2020:06:15	00:00	05:58:19	+26°17'	6.0° E
2020:06:16	00:00	06:06:11	+26°04'	6.6° E
2020:06:17	00:00	06:14:11	+25°47'	7.3° E
2020:06:18	00:00	06:22:17	+25°28'	8.1° E
2020:06:19	00:00	06:30:29	+25°07'	8.9° E
2020:06:20	00:00	06:38:45	+24°42'	9.8° E
2020:06:21	00:00	06:47:04	+24°15'	10.7° E
2020:06:22	00:00	06:55:23	+23°45'	11.6° E
2020:06:23	00:00	07:03:42	+23°12'	12.6° E
2020:06:24	00:00	07:11:58	+22°36'	13.6° E
2020:06:25	00:00	07:20:09	+21°58'	14.6° E
2020:06:26	00:00	07:28:13	+21°17'	15.6° E
2020:06:27	00:00	07:36:10	+20°34'	16.6° E
2020:06:28	00:00	07:43:58	+19°49'	17.6° E
2020:06:29	00:00	07:51:37	+19°02'	18.6° E
2020:06:30	00:00	07:59:07	+18°14'	19.6° E
2020:07:01	00:00	08:06:28	+17°24'	20.6° E
2020:07:02	00:00	08:13:41	+16°33'	21.6° E
2020:07:03	00:00	08:20:48	+15°40'	22.5° E
2020:07:04	00:00	08:27:49	+14°47'	23.5° E
2020:07:05	00:00	08:34:45	+13°52'	24.5° E
2020:07:06	00:00	08:41:39	+12°55'	25.5° E
2020:07:07	00:00	08:48:31	+11°58'	26.5° E
2020:07:08	00:00	08:55:23	+10°59'	27.5° E
2020:07:09	00:00	09:02:15	+09°59'	28.5° E
2020:07:10	00:00	09:09:10	+08°57'	29.6° E
2020:07:11	00:00	09:16:09	+07°54'	30.7° E
2020:07:12	00:00	09:23:11	+06°50'	31.8° E
2020:07:13	00:00	09:30:19	+05°44'	33.0° E
2020:07:14	00:00	09:37:33	+04°36'	34.2° E
2020:07:15	00:00	09:44:54	+03°27'	35.4° E
2020:07:16	00:00	09:52:22	+02°16'	36.7° E
2020:07:17	00:00	09:59:58	+01°04'	38.1° E
2020:07:18	00:00	10:07:43	-00°09'	39.4° E
2020:07:19	00:00	10:15:36	-01°24'	40.9° E
2020:07:20	00:00	10:23:39	-02°40'	42.3° E
2020:07:21	00:00	10:31:51	-03°57'	43.8° W
2020:07:22	00:00	10:40:12	-05°14'	45.4° E
2020:07:23	00:00	10:48:43	-06°32'	46.9° E
2020:07:24	00:00	10:57:22	-07°51'	48.5° E
2020:07:25	00:00	11:06:10	-09°09'	50.2° E
2020:07:26	00:00	11:15:06	-10°27'	51.8° E

Data	Dist(Sol) [AU]	Dist(Terra) [AU]	Magnit. [mag]
2020:06:07	0.577	1.584	8.9
2020:06:08	0.558	1.566	8.7
2020:06:09	0.540	1.548	8.4
2020:06:10	0.522	1.529	8.2
2020:06:11	0.504	1.511	7.9
2020:06:12	0.486	1.492	7.7
2020:06:13	0.469	1.472	7.4
2020:06:14	0.452	1.453	7.1
2020:06:15	0.436	1.433	6.9
2020:06:16	0.420	1.413	6.6
2020:06:17	0.405	1.392	6.3
2020:06:18	0.391	1.370	6.1
2020:06:19	0.379	1.348	5.8
2020:06:20	0.367	1.326	5.6
2020:06:21	0.357	1.302	5.4
2020:06:22	0.349	1.278	5.2
2020:06:23	0.343	1.253	5.0
2020:06:24	0.338	1.227	4.9
2020:06:25	0.336	1.201	4.8
2020:06:26	0.336	1.175	4.8
2020:06:27	0.339	1.147	4.7
2020:06:28	0.343	1.120	4.8
2020:06:29	0.350	1.093	4.8
2020:06:30	0.358	1.066	4.9
2020:07:01	0.368	1.039	5.1
2020:07:02	0.380	1.012	5.2
2020:07:03	0.393	0.986	5.4
2020:07:04	0.407	0.961	5.5
2020:07:05	0.421	0.936	5.7
2020:07:06	0.437	0.912	5.9
2020:07:07	0.454	0.888	6.1
2020:07:08	0.470	0.866	6.3
2020:07:09	0.488	0.844	6.5
2020:07:10	0.506	0.824	6.6
2020:07:11	0.523	0.804	6.8
2020:07:12	0.542	0.785	7.0
2020:07:13	0.560	0.767	7.1
2020:07:14	0.578	0.750	7.3
2020:07:15	0.597	0.734	7.5
2020:07:16	0.615	0.719	7.6
2020:07:17	0.634	0.705	7.8
2020:07:18	0.652	0.692	7.9
2020:07:19	0.671	0.680	8.1
2020:07:20	0.689	0.669	8.2
2020:07:21	0.708	0.659	8.3
2020:07:22	0.726	0.650	8.5
2020:07:23	0.744	0.642	8.6
2020:07:24	0.762	0.635	8.7
2020:07:25	0.780	0.629	8.9
2020:07:26	0.798	0.624	9.0

141P/Machholz

Data	Ora	AR [h m s]	Dec [° ']	Elongazione [°]	
2020:11:07	00:00	18:35:26	-09°59'	55.0°	E
2020:11:08	00:00	18:38:06	-10°00'	54.7°	E
2020:11:09	00:00	18:40:48	-10°02'	54.4°	E
2020:11:10	00:00	18:43:31	-10°04'	54.0°	E
2020:11:11	00:00	18:46:16	-10°06'	53.7°	E
2020:11:12	00:00	18:49:03	-10°08'	53.4°	E
2020:11:13	00:00	18:51:51	-10°10'	53.1°	E
2020:11:14	00:00	18:54:41	-10°13'	52.8°	E
2020:11:15	00:00	18:57:32	-10°16'	52.5°	E
2020:11:16	00:00	19:00:25	-10°20'	52.2°	E
2020:11:17	00:00	19:03:20	-10°24'	51.9°	E
2020:11:18	00:00	19:06:16	-10°28'	51.6°	E
2020:11:19	00:00	19:09:14	-10°33'	51.3°	E
2020:11:20	00:00	19:12:14	-10°38'	51.0°	E
2020:11:21	00:00	19:15:15	-10°44'	50.7°	E
2020:11:22	00:00	19:18:19	-10°51'	50.4°	E
2020:11:23	00:00	19:21:24	-10°58'	50.1°	E
2020:11:24	00:00	19:24:32	-11°06'	49.8°	E
2020:11:25	00:00	19:27:42	-11°15'	49.5°	E
2020:11:26	00:00	19:30:55	-11°25'	49.2°	E
2020:11:27	00:00	19:34:10	-11°35'	49.0°	E
2020:11:28	00:00	19:37:29	-11°47'	48.7°	E
2020:11:29	00:00	19:40:51	-12°00'	48.4°	E
2020:11:30	00:00	19:44:16	-12°14'	48.2°	E
2020:12:01	00:00	19:47:46	-12°29'	47.9°	E
2020:12:02	00:00	19:51:20	-12°45'	47.6°	E
2020:12:03	00:00	19:54:59	-13°03'	47.4°	E
2020:12:04	00:00	19:58:45	-13°23'	47.2°	E
2020:12:05	00:00	20:02:36	-13°44'	46.9°	E
2020:12:06	00:00	20:06:35	-14°07'	46.7°	E
2020:12:07	00:00	20:10:41	-14°32'	46.6°	E
2020:12:08	00:00	20:14:57	-14°59'	46.4°	E
2020:12:09	00:00	20:19:23	-15°29'	46.3°	E
2020:12:10	00:00	20:24:01	-16°00'	46.2°	E
2020:12:11	00:00	20:28:51	-16°34'	46.1°	E
2020:12:12	00:00	20:33:56	-17°11'	46.1°	E
2020:12:13	00:00	20:39:18	-17°51'	46.1°	E
2020:12:14	00:00	20:44:59	-18°33'	46.2°	E
2020:12:15	00:00	20:51:00	-19°19'	46.4°	E
2020:12:16	00:00	20:57:26	-20°08'	46.7°	E
2020:12:17	00:00	21:04:18	-21°01'	47.0°	E
2020:12:18	00:00	21:11:41	-21°57'	47.5°	E
2020:12:19	00:00	21:19:40	-22°57'	48.1°	E
2020:12:20	00:00	21:28:17	-24°01'	48.8°	E
2020:12:21	00:00	21:37:39	-25°09'	49.7°	E
2020:12:22	00:00	21:47:51	-26°20'	50.8°	E
2020:12:23	00:00	21:59:00	-27°34'	52.0°	E
2020:12:24	00:00	22:11:11	-28°50'	53.5°	E
2020:12:25	00:00	22:24:31	-30°09'	55.2°	E
2020:12:26	00:00	22:39:07	-31°27'	57.2°	E

Data	Ora	AR [h m s]	Dec [° ']	Elongazione [°]	
2020:12:27	00:00	22:55:03	-32°44'	59.4°	E
2020:12:28	00:00	23:12:23	-33°59'	61.9°	E
2020:12:29	00:00	23:31:07	-35°07'	64.6°	E
2020:12:30	00:00	23:51:12	-36°07'	67.5°	E
2020:12:31	00:00	00:12:28	-36°56'	70.6°	E
2021:01:01	00:00	00:34:40	-37°32'	73.9°	E
2021:01:02	00:00	00:57:27	-37°52'	77.3°	E
2021:01:03	00:00	01:20:27	-37°56'	80.8°	E
2021:01:04	00:00	01:43:13	-37°43'	84.3°	E
2021:01:05	00:00	02:05:20	-37°15'	87.7°	E
2021:01:06	00:00	02:26:30	-36°33'	91.1°	E
2021:01:07	00:00	02:46:26	-35°39'	94.3°	E
2021:01:08	00:00	03:05:01	-34°37'	97.4°	E
2021:01:09	00:00	03:22:10	-33°28'	100.3°	E
2021:01:10	00:00	03:37:53	-32°15'	103.1°	E

Data	Dist(Sol) [AU]	Dist(Terra) [AU]	Magnit. [mag]
2020:11:07	0.832	0.752	9.0
2020:11:08	0.826	0.741	8.9
2020:11:09	0.819	0.730	8.7
2020:11:10	0.813	0.718	8.6
2020:11:11	0.807	0.706	8.5
2020:11:12	0.801	0.695	8.3
2020:11:13	0.796	0.683	8.2
2020:11:14	0.791	0.671	8.1
2020:11:15	0.786	0.659	8.0
2020:11:16	0.782	0.646	7.8
2020:11:17	0.778	0.634	7.7
2020:11:18	0.774	0.621	7.6
2020:11:19	0.771	0.609	7.5
2020:11:20	0.768	0.596	7.4
2020:11:21	0.765	0.583	7.3
2020:11:22	0.763	0.571	7.3
2020:11:23	0.761	0.558	7.2
2020:11:24	0.760	0.545	7.1
2020:11:25	0.759	0.532	7.0
2020:11:26	0.758	0.519	7.0
2020:11:27	0.758	0.506	6.9
2020:11:28	0.758	0.493	6.9
2020:11:29	0.758	0.480	6.8
2020:11:30	0.759	0.467	6.8
2020:12:01	0.760	0.454	6.7
2020:12:02	0.762	0.441	6.7
2020:12:03	0.764	0.429	6.6
2020:12:04	0.766	0.416	6.6
2020:12:05	0.769	0.403	6.6
2020:12:06	0.772	0.391	6.6
2020:12:07	0.775	0.378	6.6
2020:12:08	0.779	0.366	6.6
2020:12:09	0.783	0.354	6.6
2020:12:10	0.788	0.342	6.6
2020:12:11	0.792	0.330	6.6
2020:12:12	0.797	0.318	6.6
2020:12:13	0.803	0.307	6.6
2020:12:14	0.809	0.296	6.6
2020:12:15	0.815	0.284	6.6
2020:12:16	0.821	0.274	6.6
2020:12:17	0.827	0.263	6.6
2020:12:18	0.834	0.253	6.7
2020:12:19	0.841	0.243	6.7
2020:12:20	0.848	0.233	6.7
2020:12:21	0.856	0.224	6.7
2020:12:22	0.864	0.215	6.8
2020:12:23	0.872	0.207	6.8
2020:12:24	0.880	0.199	6.8
2020:12:25	0.888	0.192	6.9
2020:12:26	0.897	0.185	6.9
2020:12:27	0.905	0.179	7.0
2020:12:28	0.914	0.174	7.0

Data	Dist(Sol) [AU]	Dist(Terra) [AU]	Magnit. [mag]
2020:12:29	0.923	0.170	7.1
2020:12:30	0.933	0.166	7.2
2020:12:31	0.942	0.163	7.3
2021:01:01	0.951	0.162	7.4
2021:01:02	0.961	0.161	7.5
2021:01:03	0.971	0.161	7.6
2021:01:04	0.980	0.162	7.8
2021:01:05	0.990	0.164	7.9
2021:01:06	1.000	0.167	8.1
2021:01:07	1.011	0.170	8.3
2021:01:08	1.021	0.175	8.5
2021:01:09	1.031	0.180	8.7
2021:01:10	1.041	0.186	8.9

Cometa 73P/Schwassmann-Wachmann

Data	Ora	AR [h m s]	Dec [° ']	Elongazione [°]
2022:06:30	00:00	11:08:04	+01°32'	69.3° E
2022:07:01	00:00	11:09:14	+00°34'	69.0° E
2022:07:02	00:00	11:10:24	-00°25'	68.7° E
2022:07:03	00:00	11:11:34	-01°27'	68.5° E
2022:07:04	00:00	11:12:45	-02°30'	68.3° E
2022:07:05	00:00	11:13:56	-03°36'	68.1° E
2022:07:06	00:00	11:15:07	-04°43'	67.9° E
2022:07:07	00:00	11:16:19	-05°53'	67.7° E
2022:07:08	00:00	11:17:31	-07°05'	67.6° E
2022:07:09	00:00	11:18:44	-08°20'	67.5° E
2022:07:10	00:00	11:19:57	-09°37'	67.5° E
2022:07:11	00:00	11:21:11	-10°57'	67.5° E
2022:07:12	00:00	11:22:26	-12°20'	67.5° E
2022:07:13	00:00	11:23:42	-13°45'	67.6° E
2022:07:14	00:00	11:24:59	-15°13'	67.7° E
2022:07:15	00:00	11:26:18	-16°45'	67.9° E
2022:07:16	00:00	11:27:38	-18°19'	68.1° E
2022:07:17	00:00	11:29:00	-19°57'	68.4° E
2022:07:18	00:00	11:30:24	-21°38'	68.8° E
2022:07:19	00:00	11:31:51	-23°22'	69.2° E
2022:07:20	00:00	11:33:21	-25°10'	69.7° E
2022:07:21	00:00	11:34:54	-27°01'	70.2° E
2022:07:22	00:00	11:36:31	-28°56'	70.8° E
2022:07:23	00:00	11:38:13	-30°53'	71.5° E
2022:07:24	00:00	11:40:00	-32°55'	72.3° E
2022:07:25	00:00	11:41:54	-34°59'	73.1° E
2022:07:26	00:00	11:43:55	-37°06'	74.0° E
2022:07:27	00:00	11:46:04	-39°17'	75.0° E
2022:07:28	00:00	11:48:23	-41°30'	76.1° E
2022:07:29	00:00	11:50:54	-43°46'	77.2° E
2022:07:30	00:00	11:53:39	-46°04'	78.4° E
2022:07:31	00:00	11:56:40	-48°24'	79.7° E

```
    Data    Dist(Sol) Dist(Terra) Magnit.
             [AU]       [AU]       [mag]
----------------------------------------
2022:06:30   0.951      0.365       9.0
2022:07:01   0.949      0.358       8.9
2022:07:02   0.948      0.352       8.9
2022:07:03   0.946      0.346       8.8
2022:07:04   0.945      0.340       8.8
2022:07:05   0.944      0.334       8.7
2022:07:06   0.944      0.327       8.7
2022:07:07   0.943      0.321       8.7
2022:07:08   0.943      0.315       8.6
2022:07:09   0.943      0.309       8.6
2022:07:10   0.943      0.303       8.5
2022:07:11   0.944      0.298       8.5
2022:07:12   0.944      0.292       8.5
2022:07:13   0.945      0.286       8.4
2022:07:14   0.947      0.281       8.4
2022:07:15   0.948      0.275       8.4
2022:07:16   0.950      0.270       8.3
2022:07:17   0.951      0.265       8.3
2022:07:18   0.953      0.260       8.3
2022:07:19   0.956      0.256       8.2
2022:07:20   0.958      0.251       8.2
2022:07:21   0.961      0.247       8.2
2022:07:22   0.964      0.243       8.2
2022:07:23   0.967      0.239       8.2
2022:07:24   0.970      0.235       8.2
2022:07:25   0.974      0.232       8.2
2022:07:26   0.978      0.228       8.1
2022:07:27   0.982      0.226       8.1
2022:07:28   0.986      0.223       8.1
2022:07:29   0.990      0.221       8.2
2022:07:30   0.995      0.219       8.2
2022:07:31   0.999      0.217       8.2
```

```
Cometa 2P/Encke

   Data        Ora        AR          Dec      Elongazione
                        [h m s]      [° ']        [ ° ]
--------------------------------------------------------
2023:09:19    00:00    09:31:36    +22°47'      38.6°  W
2023:09:20    00:00    09:39:50    +21°53'      37.4°  W
2023:09:21    00:00    09:48:01    +20°56'      36.2°  W
2023:09:22    00:00    09:56:09    +19°57'      35.1°  W
2023:09:23    00:00    10:04:13    +18°57'      33.9°  W
2023:09:24    00:00    10:12:14    +17°55'      32.7°  W
2023:09:25    00:00    10:20:10    +16°51'      31.5°  W
2023:09:26    00:00    10:28:04    +15°46'      30.2°  W
2023:09:27    00:00    10:35:54    +14°39'      29.0°  W
2023:09:28    00:00    10:43:40    +13°32'      27.8°  W
2023:09:29    00:00    10:51:24    +12°23'      26.5°  W
2023:09:30    00:00    10:59:06    +11°13'      25.3°  W
2023:10:01    00:00    11:06:46    +10°02'      24.0°  W
2023:10:02    00:00    11:14:25    +08°49'      22.8°  W
2023:10:03    00:00    11:22:03    +07°37'      21.5°  W
2023:10:04    00:00    11:29:41    +06°23'      20.2°  W
2023:10:05    00:00    11:37:20    +05°08'      18.9°  W
2023:10:06    00:00    11:45:00    +03°53'      17.6°  W
2023:10:07    00:00    11:52:43    +02°38'      16.3°  W
2023:10:08    00:00    12:00:28    +01°22'      15.0°  W
2023:10:09    00:00    12:08:16    +00°06'      13.6°  W
2023:10:10    00:00    12:16:07    -01°11'      12.3°  W
2023:10:11    00:00    12:24:01    -02°26'      10.9°  W
2023:10:12    00:00    12:31:58    -03°42'       9.6°  W
2023:10:13    00:00    12:39:57    -04°56'       8.3°  W
2023:10:14    00:00    12:47:57    -06°10'       7.0°  W
2023:10:15    00:00    12:55:58    -07°21'       5.8°  W
2023:10:16    00:00    13:03:58    -08°31'       4.7°  W
2023:10:17    00:00    13:11:57    -09°40'       3.6°  W
2023:10:18    00:00    13:19:54    -10°46'       2.9°  W
2023:10:19    00:00    13:27:47    -11°49'       2.5°  W
2023:10:20    00:00    13:35:37    -12°51'       2.7°  W
2023:10:21    00:00    13:43:23    -13°50'       3.3°  E
2023:10:22    00:00    13:51:05    -14°46'       4.1°  E
2023:10:23    00:00    13:58:42    -15°40'       5.0°  E
2023:10:24    00:00    14:06:14    -16°31'       5.9°  E
2023:10:25    00:00    14:13:41    -17°21'       6.8°  E
2023:10:26    00:00    14:21:03    -18°07'       7.7°  E
2023:10:27    00:00    14:28:20    -18°52'       8.5°  E
2023:10:28    00:00    14:35:31    -19°34'       9.3°  E
2023:10:29    00:00    14:42:37    -20°14'      10.1°  E
2023:10:30    00:00    14:49:37    -20°52'      10.8°  E
2023:10:31    00:00    14:56:32    -21°28'      11.5°  E
```

Data	Dist(Sol) [AU]	Dist(Terra) [AU]	Magnit. [mag]
2023:09:19	0.667	1.014	8.9
2023:09:20	0.648	1.016	8.7
2023:09:21	0.630	1.020	8.5
2023:09:22	0.611	1.024	8.3
2023:09:23	0.593	1.030	8.2
2023:09:24	0.574	1.036	8.0
2023:09:25	0.556	1.043	7.8
2023:09:26	0.538	1.051	7.6
2023:09:27	0.520	1.060	7.4
2023:09:28	0.502	1.070	7.2
2023:09:29	0.484	1.081	6.9
2023:09:30	0.467	1.093	6.7
2023:10:01	0.450	1.105	6.5
2023:10:02	0.434	1.119	6.3
2023:10:03	0.418	1.133	6.1
2023:10:04	0.403	1.148	5.9
2023:10:05	0.390	1.163	5.7
2023:10:06	0.377	1.179	5.5
2023:10:07	0.366	1.195	5.3
2023:10:08	0.356	1.211	5.2
2023:10:09	0.348	1.228	5.1
2023:10:10	0.342	1.244	5.0
2023:10:11	0.338	1.260	4.9
2023:10:12	0.336	1.276	4.9
2023:10:13	0.337	1.292	5.0
2023:10:14	0.339	1.307	5.0
2023:10:15	0.344	1.321	5.2
2023:10:16	0.351	1.335	5.3
2023:10:17	0.359	1.348	5.5
2023:10:18	0.369	1.361	5.7
2023:10:19	0.381	1.374	5.9
2023:10:20	0.394	1.386	6.1
2023:10:21	0.408	1.398	6.4
2023:10:22	0.423	1.410	6.6
2023:10:23	0.439	1.422	6.9
2023:10:24	0.456	1.434	7.2
2023:10:25	0.473	1.445	7.4
2023:10:26	0.490	1.457	7.7
2023:10:27	0.508	1.469	7.9
2023:10:28	0.526	1.481	8.2
2023:10:29	0.544	1.493	8.4
2023:10:30	0.562	1.506	8.6
2023:10:31	0.581	1.519	8.9

Cometa 2P/Encke

Data	Ora	AR [h m s]	Dec [° ']	Elongazione [°]	
2027:01:01	00:00	21:40:32	+00°19'	48.6°	E
2027:01:02	00:00	21:38:26	-00°06'	47.0°	E
2027:01:03	00:00	21:36:10	-00°31'	45.3°	E
2027:01:04	00:00	21:33:46	-00°59'	43.6°	E
2027:01:05	00:00	21:31:11	-01°28'	41.8°	E
2027:01:06	00:00	21:28:24	-01°59'	40.0°	E
2027:01:07	00:00	21:25:26	-02°32'	38.1°	E
2027:01:08	00:00	21:22:14	-03°07'	36.2°	E
2027:01:09	00:00	21:18:48	-03°45'	34.2°	E
2027:01:10	00:00	21:15:06	-04°26'	32.1°	E
2027:01:11	00:00	21:11:08	-05°09'	29.9°	E
2027:01:12	00:00	21:06:53	-05°55'	27.6°	E
2027:01:13	00:00	21:02:20	-06°44'	25.3°	E
2027:01:14	00:00	20:57:29	-07°37'	22.9°	E
2027:01:15	00:00	20:52:20	-08°32'	20.4°	E
2027:01:16	00:00	20:46:53	-09°30'	17.8°	E
2027:01:17	00:00	20:41:11	-10°32'	15.2°	E
2027:01:18	00:00	20:35:15	-11°35'	12.6°	E
2027:01:19	00:00	20:29:07	-12°41'	9.9°	E
2027:01:20	00:00	20:22:53	-13°48'	7.4°	E
2027:01:21	00:00	20:16:38	-14°55'	5.2°	E
2027:01:22	00:00	20:10:28	-16°03'	3.9°	W
2027:01:23	00:00	20:04:29	-17°09'	4.4°	W
2027:01:24	00:00	19:58:50	-18°13'	6.1°	W
2027:01:25	00:00	19:53:39	-19°14'	8.2°	W
2027:01:26	00:00	19:49:01	-20°11'	10.4°	W
2027:01:27	00:00	19:45:03	-21°04'	12.4°	W
2027:01:28	00:00	19:41:50	-21°52'	14.3°	W
2027:01:29	00:00	19:39:22	-22°35'	16.0°	W
2027:01:30	00:00	19:37:42	-23°12'	17.6°	W
2027:01:31	00:00	19:36:46	-23°45'	18.9°	W
2027:02:01	00:00	19:36:32	-24°13'	20.1°	W
2027:02:02	00:00	19:36:57	-24°37'	21.2°	W
2027:02:03	00:00	19:37:54	-24°57'	22.1°	W
2027:02:04	00:00	19:39:20	-25°13'	22.8°	W
2027:02:05	00:00	19:41:10	-25°26'	23.5°	W
2027:02:06	00:00	19:43:20	-25°36'	24.1°	W
2027:02:07	00:00	19:45:45	-25°43'	24.6°	W
2027:02:08	00:00	19:48:23	-25°48'	25.1°	W
2027:02:09	00:00	19:51:12	-25°52'	25.5°	W
2027:02:10	00:00	19:54:07	-25°53'	25.9°	W
2027:02:11	00:00	19:57:08	-25°53'	26.3°	W
2027:02:12	00:00	20:00:13	-25°52'	26.6°	W
2027:02:13	00:00	20:03:20	-25°50'	26.9°	W
2027:02:14	00:00	20:06:29	-25°47'	27.2°	W
2027:02:15	00:00	20:09:39	-25°42'	27.6°	W
2027:02:16	00:00	20:12:48	-25°37'	27.9°	W
2027:02:17	00:00	20:15:56	-25°32'	28.2°	W

Data	Dist(Sol) [AU]	Dist(Terra) [AU]	Magnit. [mag]
2027:01:01	0.738	0.646	8.6
2027:01:02	0.720	0.641	8.4
2027:01:03	0.701	0.636	8.2
2027:01:04	0.683	0.630	8.0
2027:01:05	0.665	0.625	7.8
2027:01:06	0.646	0.619	7.6
2027:01:07	0.628	0.614	7.4
2027:01:08	0.609	0.609	7.2
2027:01:09	0.591	0.604	7.0
2027:01:10	0.572	0.599	6.8
2027:01:11	0.554	0.595	6.5
2027:01:12	0.536	0.591	6.3
2027:01:13	0.517	0.587	6.1
2027:01:14	0.500	0.585	5.8
2027:01:15	0.482	0.583	5.6
2027:01:16	0.465	0.582	5.3
2027:01:17	0.448	0.583	5.1
2027:01:18	0.432	0.585	4.9
2027:01:19	0.416	0.589	4.6
2027:01:20	0.402	0.594	4.4
2027:01:21	0.388	0.602	4.2
2027:01:22	0.376	0.612	4.1
2027:01:23	0.365	0.625	3.9
2027:01:24	0.355	0.639	3.8
2027:01:25	0.347	0.657	3.7
2027:01:26	0.342	0.676	3.7
2027:01:27	0.338	0.698	3.6
2027:01:28	0.336	0.722	3.7
2027:01:29	0.337	0.748	3.8
2027:01:30	0.340	0.775	3.9
2027:01:31	0.345	0.803	4.1
2027:02:01	0.352	0.832	4.3
2027:02:02	0.360	0.861	4.5
2027:02:03	0.371	0.891	4.8
2027:02:04	0.383	0.920	5.1
2027:02:05	0.396	0.949	5.3
2027:02:06	0.410	0.978	5.6
2027:02:07	0.425	1.006	5.9
2027:02:08	0.441	1.033	6.2
2027:02:09	0.458	1.060	6.5
2027:02:10	0.475	1.086	6.8
2027:02:11	0.492	1.112	7.1
2027:02:12	0.510	1.136	7.4
2027:02:13	0.528	1.161	7.7
2027:02:14	0.546	1.184	7.9
2027:02:15	0.564	1.207	8.2
2027:02:16	0.583	1.229	8.4
2027:02:17	0.601	1.251	8.7
2027:02:18	0.620	1.272	8.9

EVOLUZIONE DEGLI ELEMENTI ORBITALI
ORBITAL ELEMENTS EVOLUTION

2P/ENCKE

Date	e	i	w	Node	a	M0
2010/01/01	0.848260822	11.7842121	186.5239447	334.5714593	2.215396347	294.9860916
2011/01/01	0.848311805	11.7827967	186.5500697	334.5668700	2.214414767	44.1924299
2012/01/01	0.848137595	11.7770558	186.5475822	334.5717522	2.214274819	153.4907039
2012/12/31	0.848190910	11.7786757	186.5385765	334.5738858	2.214336290	262.7999471
2014/01/01	0.848232376	11.7786607	186.5367212	334.5725152	2.214746716	12.0543871
2015/01/01	0.848303064	11.7812188	186.5420815	334.5687930	2.215114165	121.2348245
2016/01/01	0.848356059	11.7813558	186.5490401	334.5682970	2.215067432	230.3856309
2016/12/31	0.848333701	11.7782759	186.5617991	334.5608175	2.214737324	339.5677178
2018/01/01	0.848275785	11.7766681	186.5636098	334.5622010	2.214530128	88.8059526
2019/01/01	0.848115277	11.7699878	186.5659382	334.5618545	2.214707553	198.0497046
2020/01/01	0.848000156	11.7649117	186.5625203	334.5542177	2.215132861	307.2879751
2020/12/31	0.848091279	11.7693927	186.5764597	334.5406144	2.216062738	56.4389456
2022/01/01	0.849737219	11.7601267	186.7050834	334.5033737	2.218541563	164.5684187
2023/01/01	0.847280263	11.3654048	187.2102429	334.1074536	2.219586121	272.3362452
2024/01/01	0.846941181	11.3368753	187.2871840	334.0191295	2.218656627	21.1787684
2024/12/31	0.847040366	11.3416412	187.2801195	334.0147714	2.218238888	130.1437420
2026/01/01	0.847232806	11.3469392	187.2794268	334.0181066	2.217984701	239.1032958
2027/01/01	0.847334054	11.3481550	187.2876277	334.0208383	2.217744849	348.0877082
2028/01/01	0.847309962	11.3476190	187.2965885	334.0176296	2.217974449	97.0712057
2028/12/31	0.847209108	11.3418909	187.3063493	334.0156177	2.218217410	206.0058523
2030/01/01	0.847076662	11.3349155	187.3161894	334.0021555	2.218248543	314.9499637
2031/01/01	0.847060088	11.3353795	187.3229023	333.9908644	2.217812640	63.9318713
2032/01/01	0.847096414	11.3377177	187.3122987	333.9905453	2.217431164	172.9968103
2032/12/31	0.847328118	11.3430976	187.3033287	333.9985979	2.217407717	282.0815437
2034/01/01	0.847429975	11.3436772	187.3186294	333.9948373	2.217610480	31.1093200
2035/01/01	0.846938994	11.3253239	187.3423446	334.0069643	2.218602876	140.0740032
2036/01/01	0.846631409	11.3105835	187.3394370	333.9968417	2.219105408	249.0719524
2036/12/31	0.846594257	11.3066405	187.3373650	333.9813288	2.219927979	357.9780463
2038/01/01	0.846692537	11.3103201	187.3398652	333.9747798	2.220116533	106.7944446
2039/01/01	0.846817069	11.3133385	187.3441696	333.9754147	2.220040576	215.5795512
2040/01/01	0.846865457	11.3125593	187.3576272	333.9731266	2.219571970	324.3906901
2040/12/31	0.846821578	11.3116536	187.3653710	333.9709671	2.219378646	73.2675080
2042/01/01	0.846643605	11.3039120	187.3735405	333.9713705	2.219707913	182.1202614
2043/01/01	0.846467944	11.2964519	187.3766658	333.9605114	2.220006060	290.9718523
2044/01/01	0.846475877	11.2965913	187.3792844	333.9472404	2.220390519	39.7921383
2044/12/31	0.847480820	11.3347596	187.3566750	333.9260539	2.219369439	148.5869392
2046/01/01	0.850221785	11.2966066	187.5477565	333.8521901	2.214144888	257.1406793
2047/01/01	0.850272618	11.2532922	187.7292415	333.7331811	2.210458748	6.3740588
2048/01/01	0.850297227	11.2557846	187.7212548	333.7306412	2.209864301	115.9649518
2048/12/31	0.850482928	11.2633229	187.7108015	333.7353834	2.209596761	225.5937478
2050/01/01	0.850624026	11.2670377	187.7105862	333.7434596	2.209522531	335.2098881
2051/01/01	0.850664598	11.2681685	187.7156341	333.7411976	2.210011785	84.7866423
2052/01/01	0.850672084	11.2664986	187.7244442	333.7405343	2.210219418	194.2989966
2052/12/31	0.850628850	11.2616486	187.7387221	333.7314995	2.209987118	303.8226853
2054/01/01	0.850589007	11.2609047	187.7423427	333.7258198	2.209424773	53.4172055
2055/01/01	0.850456724	11.2556533	187.7432109	333.7274120	2.209457593	163.0542893
2056/01/01	0.850395318	11.2515220	187.7406364	333.7230284	2.209695815	272.6988672
2056/12/31	0.850433361	11.2511345	187.7431992	333.7199243	2.210135097	22.2923323
2058/01/01	0.849615049	11.2139704	187.8355642	333.7327661	2.214356946	131.4781611
2059/01/01	0.848024712	11.1230022	187.9024910	333.6708312	2.216541629	240.6181811
2060/01/01	0.847644604	11.1004185	187.9245899	333.6097948	2.217502855	349.6917447
2060/12/31	0.847734306	11.1043210	187.9221650	333.6029912	2.217327823	98.7151134
2062/01/01	0.847941449	11.1115042	187.9189413	333.6052315	2.217100235	207.7386069
2063/01/01	0.848057581	11.1133563	187.9278028	333.6084182	2.216691788	316.7739212
2064/01/01	0.848049055	11.1140978	187.9444319	333.6001131	2.216694046	65.8517939
2064/12/31	0.847933186	11.1086754	187.9537930	333.6007678	2.217047198	174.8863384
2066/01/01	0.847777140	11.1011028	187.9628125	333.5895313	2.217183558	283.9138734
2067/01/01	0.847751473	11.1001580	187.9664098	333.5781720	2.217168603	32.9526587
2068/01/01	0.848005155	11.1118112	187.9488999	333.5698781	2.216348784	142.0691755
2068/12/31	0.848822928	11.1281984	187.9418417	333.5847214	2.215304492	251.2417065
2070/01/01	0.849214577	11.1248820	188.0135449	333.5608007	2.213720852	0.4412706
2071/01/01	0.849081112	11.1212524	188.0139750	333.5654418	2.213458429	109.7746647
2072/01/01	0.849085897	11.1215922	188.0043317	333.5668118	2.213349935	219.1581970
2072/12/31	0.849163515	11.1233044	187.9976029	333.5704801	2.213675921	328.5078021
2074/01/01	0.849209672	11.1250649	188.0057678	333.5619913	2.214029127	77.7866879
2075/01/01	0.849280173	11.1265062	188.0104938	333.5615431	2.214123937	187.0177021
2076/01/01	0.849268143	11.1233830	188.0223109	333.5559809	2.213908811	296.2533339
2076/12/31	0.849241429	11.1224531	188.0308792	333.5511068	2.213467429	45.5496354
2078/01/01	0.849115043	11.1176764	188.0341692	333.5533349	2.213577677	154.8706604
2079/01/01	0.848962623	11.1114957	188.0341274	333.5466018	2.213872427	264.1936957
2080/01/01	0.848957479	11.1102457	188.0323415	333.5393294	2.214506055	13.4758943
2080/12/31	0.849751465	11.1363299	188.0837402	333.5135487	2.217301772	122.2993859
2082/01/01	0.847952398	10.7568111	188.5979290	333.1921045	2.221100515	229.6977806
2083/01/01	0.846842791	10.6600566	188.7897229	332.9670435	2.220775482	338.2852495
2084/01/01	0.846862439	10.6621213	188.7889092	332.9595324	2.220092171	87.1011683
2084/12/31	0.847045285	10.6677960	188.7836269	332.9606665	2.219846082	195.9357899
2086/01/01	0.847189395	10.6699881	188.7900851	332.9648014	2.219513097	304.7746994
2087/01/01	0.847208193	10.6697912	188.7989291	332.9644399	2.219642051	53.6401747
2088/01/01	0.847143732	10.6662517	188.8084217	332.9648765	2.220021898	162.4502913
2088/12/31	0.846999568	10.6589962	188.8193987	332.9546309	2.220134985	271.2514821
2090/01/01	0.846938525	10.6564221	188.8273007	332.9409106	2.219978537	20.0721976
2091/01/01	0.846967644	10.6593330	188.8181683	332.9369170	2.219345401	128.9724218
2092/01/01	0.847137978	10.6640941	188.8043623	332.9420792	2.219176426	237.9305870
2092/12/31	0.847396278	10.6669717	188.8143611	332.9489043	2.219004316	346.8505868
2094/01/01	0.847172769	10.6598738	188.8313746	332.9574456	2.219605625	95.7337170

113

Date	e	i	w	Node	a	M0
2095/01/01	0.846751835	10.6423584	188.8363590	332.9533502	2.220014745	204.6443418
2096/01/01	0.846689673	10.6379846	188.8304690	332.9441113	2.220624621	313.5148394
2096/12/31	0.846718062	10.6388247	188.8370285	332.9324546	2.220977257	62.2858408
2098/01/01	0.846845469	10.6424194	188.8389750	332.9313726	2.221043062	171.0097293
2099/01/01	0.846924108	10.6423974	188.8490997	332.9308034	2.220749104	279.7333335
2100/01/01	0.846919953	10.6415692	188.8660421	332.9221280	2.220441510	28.5160373
2101/01/01	0.846794878	10.6365186	188.8743144	332.9254633	2.220786554	137.2932611

46P/WIRTANEN

Date	e	i	w	Node	a	M0
2010/01/01	0.658265645	11.7431410	356.3072759	82.1601977	3.090932877	126.6917525
2011/01/01	0.658707116	11.7500852	356.2990526	82.1549745	3.089976164	192.9815826
2012/01/01	0.659115609	11.7558721	356.3116634	82.1595746	3.089064017	259.2588714
2012/12/31	0.659279676	11.7573598	356.3388639	82.1639579	3.087960694	325.5533941
2014/01/01	0.659264148	11.7574618	356.3351370	82.1625245	3.087623643	31.9021802
2015/01/01	0.659077000	11.7563163	356.3505714	82.1676175	3.088088187	98.2435913
2016/01/01	0.658775987	11.7526513	356.3584550	82.1732174	3.088753075	164.5596966
2016/12/31	0.658550570	11.7483474	356.3506989	82.1731470	3.089246203	230.8890935
2018/01/01	0.658490962	11.7459864	356.3367922	82.1690457	3.089960813	297.2163194
2019/01/01	0.658794985	11.7461764	356.3536437	82.1592399	3.093054194	3.4888729
2020/01/01	0.658558038	11.7450793	356.3635379	82.1565968	3.092012770	69.7105266
2020/12/31	0.658257443	11.7435144	356.3525864	82.1609199	3.091748246	135.9946422
2022/01/01	0.658364530	11.7453741	356.3295512	82.1606298	3.091385687	202.3066232
2023/01/01	0.658639365	11.7488758	356.3222058	82.1646456	3.091213409	268.5869987
2024/01/01	0.658817168	11.7497864	356.3255288	82.1672775	3.091451270	334.8401377
2024/12/31	0.658781701	11.7502188	356.3261999	82.1641080	3.091615237	41.0671317
2026/01/01	0.658642789	11.7493512	356.3543775	82.1670752	3.092770820	107.2343263
2027/01/01	0.658345427	11.7454691	356.3669923	82.1719280	3.093535058	173.3763746
2028/01/01	0.658005775	11.7399196	356.3625313	82.1704243	3.094205408	239.5294160
2028/12/31	0.657808402	11.7362822	356.3482303	82.1622562	3.094994013	305.6781839
2030/01/01	0.657744468	11.7360678	356.3384582	82.1565510	3.095104331	11.7937032
2031/01/01	0.657449328	11.7395210	356.4480470	82.1362941	3.098373921	77.7895400
2032/01/01	0.655537775	11.7341437	356.5597306	82.1429626	3.103191518	143.6535194
2032/12/31	0.654490452	11.7199283	356.5222063	82.1496168	3.105117372	209.6094879
2034/01/01	0.654239204	11.7131039	356.4817063	82.1414712	3.106756734	275.4934390
2035/01/01	0.654228273	11.7095567	356.4495098	82.1200745	3.109104200	341.2634358
2036/01/01	0.654197555	11.7103378	356.4639130	82.1118085	3.109223741	46.9244570
2036/12/31	0.654168152	11.7111297	356.4960122	82.1082500	3.110436739	112.5107605
2038/01/01	0.653960527	11.7088075	356.5154435	82.1106538	3.111128841	178.0645732
2039/01/01	0.653617463	11.7032637	356.5204142	82.1081805	3.111664138	243.6250949
2040/01/01	0.653370852	11.6989611	356.5162605	82.0972282	3.112048591	309.1962396
2040/12/31	0.653275490	11.6985943	356.5168033	82.0896402	3.111797640	14.7690250
2042/01/01	0.653896509	11.7281116	356.7654331	81.9400759	3.118326964	80.0260417
2043/01/01	0.645622358	12.3420229	359.1892563	80.8545242	3.166489104	141.5656489
2044/01/01	0.625972811	12.2027073	359.5765403	80.8427102	3.208188926	204.1105652
2044/12/31	0.622410439	12.1077777	359.4854075	80.7174325	3.217629816	266.8643743
2046/01/01	0.621364905	12.0772571	359.4714875	80.5773861	3.221719962	329.2442897
2047/01/01	0.621214528	12.0775828	359.4978440	80.5576332	3.220820667	31.5060796
2048/01/01	0.621334915	12.0795928	359.5288559	80.5471327	3.222057014	93.7162451
2048/12/31	0.621294282	12.0800049	359.5561969	80.5460570	3.222771491	155.8763644
2050/01/01	0.621012185	12.0753911	359.5766223	80.5450611	3.223280678	218.0213536
2051/01/01	0.620701670	12.0693659	359.5889766	80.5334911	3.223417520	280.1829026
2052/01/01	0.620518096	12.0669720	359.5963605	80.5183532	3.223131906	342.3734590
2052/12/31	0.620671384	12.0688566	359.6487817	80.4916908	3.224705010	44.5438302
2054/01/01	0.623664636	12.2010641	0.0476982	80.0884698	3.227600286	106.2013107
2055/01/01	0.548400391	16.4579388	14.0917987	77.3843782	3.569900459	137.9822867
2056/01/01	0.480096847	14.1977628	15.8126728	76.8056847	3.733035869	189.7764642
2056/12/31	0.475074077	14.0310064	15.7747608	76.4394857	3.749351165	240.1412553
2058/01/01	0.473427839	13.9861853	15.6732393	76.1855517	3.760259362	290.0054988
2059/01/01	0.472645060	13.9883413	15.6106041	76.0112005	3.764913379	339.4335936
2060/01/01	0.472178110	13.9912198	15.6344894	76.0053368	3.761928782	28.7392266
2060/12/31	0.472211102	13.9909647	15.6971255	76.0059381	3.764231683	77.9893985
2062/01/01	0.471985913	13.9878209	15.7499195	76.0101457	3.765996988	127.1881264
2063/01/01	0.471533070	13.9813826	15.7835415	76.0082778	3.767431013	176.3811696
2064/01/01	0.471050622	13.9750727	15.7917912	75.9966251	3.768648377	225.5867256
2064/12/31	0.470696400	13.9715505	15.7698600	75.9818392	3.769966973	274.8088949
2066/01/01	0.470580923	13.9709308	15.7330810	75.9718305	3.771363728	324.0109826
2067/01/01	0.470545148	13.9710987	15.7113267	75.9702562	3.771443996	13.1697812
2068/01/01	0.470362757	13.9716488	15.6730799	75.9647315	3.769221751	62.3910202
2068/12/31	0.470402417	13.9744497	15.6213034	75.9590438	3.767324603	111.6901029
2070/01/01	0.470841856	13.9807828	15.5727018	75.9582054	3.765545883	161.0160856
2071/01/01	0.471430861	13.9867746	15.5605542	75.9662369	3.764014241	210.3158102
2072/01/01	0.471921053	13.9892660	15.5953048	75.9740336	3.762306795	259.5870402
2072/12/31	0.472107409	13.9891621	15.6525261	75.9723372	3.760621393	308.8827764
2074/01/01	0.472247141	13.9888918	15.6771488	75.9694477	3.760886237	358.2420904
2075/01/01	0.472399056	13.9876029	15.6329509	75.9306574	3.759970599	47.6385764
2076/01/01	0.472344655	13.9928183	15.5645395	75.9115185	3.756847764	97.1665380
2076/12/31	0.472427105	13.9963524	15.5107962	75.9089779	3.755425037	146.7292091
2078/01/01	0.472975382	13.9988500	15.4581530	75.9111120	3.753840282	196.3292061
2079/01/01	0.474283401	13.9961137	15.4830830	75.9023391	3.750991548	245.9230442
2080/01/01	0.475604532	13.9875210	15.7152580	75.8419067	3.745625467	295.4248814
2080/12/31	0.475774373	13.9875802	15.8429587	75.8124629	3.743005695	345.0555649
2082/01/01	0.476030903	13.9868056	15.8569645	75.8052999	3.744694299	34.7404345
2083/01/01	0.476299372	13.9897207	15.8715857	75.7914711	3.745408320	84.3752504

114

Date	e	i	w	Node	a	M0
2084/01/01	0.476543885	13.9926030	15.8923001	75.7879220	3.745663971	133.9741274
2084/12/31	0.476595333	13.9912951	15.9272927	75.7868172	3.745847489	183.5461006
2086/01/01	0.476467591	13.9871636	15.9628198	75.7777060	3.745827635	233.1306629
2087/01/01	0.476221130	13.9834233	15.9851749	75.7590354	3.745838695	282.7477315
2088/01/01	0.476022659	13.9831464	15.9968433	75.7425411	3.745363691	332.3877497
2088/12/31	0.475832202	13.9840802	16.0077148	75.7399774	3.744181364	22.0600798
2090/01/01	0.475774260	13.9884085	16.2162558	75.7153258	3.750477884	71.5319206
2091/01/01	0.473013718	13.9805086	16.6249578	75.7184975	3.763992562	120.6008748
2092/01/01	0.470371810	13.9466788	16.6639960	75.7151374	3.770214759	169.9402667
2092/12/31	0.469520855	13.9314522	16.5982288	75.6910125	3.772888610	219.2867326
2094/01/01	0.469464739	13.9272779	16.5342734	75.6750641	3.774942128	268.5457973
2095/01/01	0.469647283	13.9262128	16.4599798	75.6558463	3.778574106	317.7288251
2096/01/01	0.469573965	13.9287004	16.3664212	75.6438414	3.779979950	6.7681745
2096/12/31	0.469358792	13.9286629	16.3782343	75.6428683	3.779120114	55.7726084
2098/01/01	0.469110671	13.9262247	16.4157382	75.6489122	3.780360578	104.7328528
2099/01/01	0.468681757	13.9209363	16.4375741	75.6509080	3.781580133	153.6863819
2100/01/01	0.468206302	13.9155575	16.4310843	75.6450267	3.782820940	202.6604659
2101/01/01	0.467867051	13.9123653	16.3986393	75.6357511	3.784258857	251.6387646

73P/SCHWASSMANN-WACHMANN 3

Date	e	i	w	Node	a	M0
2010/01/01	0.692575123	11.3840971	198.8851561	69.8845817	3.062204345	239.9519120
2011/01/01	0.692284050	11.3799141	198.8843457	69.8626813	3.062903302	307.1247506
2012/01/01	0.692174247	11.3798317	198.8814974	69.8557022	3.062626534	14.2828200
2012/12/31	0.691845919	11.3795154	199.0999670	69.8489181	3.072743733	81.0330787
2014/01/01	0.687825445	11.2931888	199.3950891	69.8364735	3.084507059	147.3718181
2015/01/01	0.686204176	11.2524119	199.4299423	69.7681510	3.087507344	213.9169502
2016/01/01	0.685657984	11.2388862	199.4198047	69.7155861	3.089814901	280.3547827
2016/12/31	0.685446278	11.2357130	199.4103897	69.6737742	3.091330687	346.6420899
2018/01/01	0.685508536	11.2363736	199.4203596	69.6702844	3.091957090	52.8562675
2019/01/01	0.685612124	11.2371813	199.4320175	69.6689987	3.092550148	119.0146493
2020/01/01	0.685635846	11.2349588	199.4523969	69.6657545	3.092772907	185.1357258
2020/12/31	0.685504919	11.2301241	199.4802090	69.6511009	3.092679907	251.2607090
2022/01/01	0.685304190	11.2266011	199.5037679	69.6276174	3.092276747	317.4181330
2023/01/01	0.685322040	11.2270194	199.5238624	69.6160542	3.092589220	23.6141656
2024/01/01	0.687138806	11.2712689	199.4579633	69.5369424	3.090417262	89.7779837
2024/12/31	0.705146861	11.0775390	199.0319647	69.2710038	3.053369994	157.0302045
2026/01/01	0.705328952	6.3940434	213.4484105	53.7391094	3.052210289	227.5096166
2027/01/01	0.700143965	6.2661716	214.7516112	52.2304429	3.061022088	294.7498191
2028/01/01	0.699579937	6.2642598	214.8213249	52.1113421	3.059081296	1.9403758
2028/12/31	0.699728285	6.2650407	214.8219640	52.1057564	3.059641549	69.1982522
2030/01/01	0.700030447	6.2650925	214.8172562	52.1051251	3.059481936	136.4312564
2031/01/01	0.700371672	6.2631884	214.8415192	52.0932115	3.058930034	203.6491434
2032/01/01	0.700525998	6.2608205	214.8918557	52.0670155	3.058065888	270.8849580
2032/12/31	0.700484844	6.2602050	214.9374134	52.0441985	3.056661810	338.1792564
2034/01/01	0.700447304	6.2612299	214.9277069	52.0351365	3.055212723	45.5760823
2035/01/01	0.700442372	6.2623990	214.8895970	52.0352700	3.053430248	113.1089506
2036/01/01	0.700845449	6.2622536	214.8471843	52.0336135	3.052065614	180.7459745
2036/12/31	0.702746858	6.2495967	214.9931064	51.9075383	3.048987077	248.4210961
2038/01/01	0.704344879	6.2293735	215.5788126	51.5320350	3.040678377	315.9503343
2039/01/01	0.704342241	6.2296414	215.6256773	51.5103770	3.039325534	23.8596929
2040/01/01	0.704461727	6.2312083	215.6022740	51.5092107	3.038409577	91.8364114
2040/12/31	0.704777604	6.2328965	215.5831945	51.5149831	3.037754673	159.8251171
2042/01/01	0.705139384	6.2334095	215.5877732	51.5180672	3.037045206	227.8106526
2043/01/01	0.705350584	6.2329452	215.6187374	51.5102885	3.036152428	295.8086459
2044/01/01	0.705414564	6.2328709	215.6410852	51.5040197	3.035671598	3.8578604
2044/12/31	0.705279483	6.2319466	215.6491720	51.5038786	3.035444275	71.9351927
2046/01/01	0.704943029	6.2290976	215.6718157	51.4963211	3.036060281	140.0019436
2047/01/01	0.704626735	6.2273430	215.6722483	51.4856119	3.036622824	208.0902348
2048/01/01	0.704628040	6.2271042	215.6534445	51.4828728	3.037375903	276.1872163
2048/12/31	0.704819101	6.2270111	215.6618044	51.4779688	3.037890848	344.2122200
2050/01/01	0.704779619	6.2270077	215.6527747	51.4771485	3.037468669	52.2150544
2051/01/01	0.704733747	6.2274531	215.6353047	51.4781301	3.036642597	120.2854806
2052/01/01	0.704926416	6.2288231	215.6132552	51.4850891	3.036109150	188.3733752
2052/12/31	0.705204223	6.2297191	215.6066020	51.4931449	3.035721118	256.4522683
2054/01/01	0.705372041	6.2297869	215.6167602	51.4935104	3.035558876	324.5246163
2055/01/01	0.705502673	6.2294100	215.6419970	51.4746227	3.037247577	32.5650846
2056/01/01	0.705428599	6.2275379	215.6660068	51.4735862	3.038267402	100.5248074
2056/12/31	0.705136152	6.2238749	215.6995510	51.4580636	3.039081140	168.4492031
2058/01/01	0.704756982	6.2210237	215.7175889	51.4351086	3.039781277	236.3823613
2059/01/01	0.704498311	6.2201711	215.7014502	51.4220696	3.040879389	304.3221221
2060/01/01	0.704472190	6.2198511	215.6951777	51.4187442	3.041391171	12.2076360
2060/12/31	0.704074218	6.2133484	215.7730529	51.4214949	3.044266617	79.9847940
2062/01/01	0.702990807	6.1976497	215.8635011	51.3755628	3.046214270	147.7492096
2063/01/01	0.702573575	6.1900832	215.8964003	51.3255217	3.046966222	215.5184114
2064/01/01	0.702494616	6.1875495	215.9066671	51.2940960	3.047741256	283.2476892
2064/12/31	0.702474388	6.1875624	215.8964433	51.2789535	3.048789809	350.9147358
2066/01/01	0.702537640	6.1874633	215.9057544	51.2790011	3.049622459	58.5119572
2067/01/01	0.702572465	6.1860770	215.9229789	51.2757422	3.050278353	126.0511983
2068/01/01	0.702472288	6.1830696	215.9560731	51.2584029	3.050634164	193.5584961
2068/12/31	0.702235525	6.1803696	215.9874544	51.2302747	3.050802619	261.0769015
2070/01/01	0.702017544	6.1795548	215.9950908	51.2083364	3.050925965	328.6181482

115

Date	e	i	w	Node	a	M0
2071/01/01	0.702173304	6.1787207	216.0180272	51.1851229	3.052801203	36.1349893
2072/01/01	0.708138289	6.0452424	216.2981934	50.9018375	3.060607638	102.4552525
2072/12/31	0.689069867	2.8061296	252.5232088	16.3312171	3.141098132	163.6693883
2074/01/01	0.682648127	2.7767089	254.5440495	14.1886314	3.153682341	228.4344686
2075/01/01	0.680937531	2.7775491	254.6141932	13.9760828	3.158979588	292.7639454
2076/01/01	0.680462295	2.7777834	254.5799191	13.9648292	3.158905545	356.8623202
2076/12/31	0.680583116	2.7774171	254.5857847	13.9623288	3.159790937	60.9516056
2078/01/01	0.680734040	2.7764747	254.6144074	13.9393271	3.160143541	124.9969566
2079/01/01	0.680834391	2.7756130	254.6758955	13.8934113	3.160161187	189.0111283
2080/01/01	0.680765773	2.7754463	254.7355626	13.8499131	3.159894822	253.0295107
2080/12/31	0.680600534	2.7757319	254.7652041	13.8286117	3.159150646	317.0809168
2082/01/01	0.680610251	2.7757412	254.7748105	13.8280336	3.159216107	21.1888661
2083/01/01	0.681703342	2.7693114	254.7322096	13.7320312	3.155896189	85.3777340
2084/01/01	0.686550655	2.7323910	255.8627899	12.3080232	3.143239457	150.1173661
2084/12/31	0.696321253	6.4198317	328.5703373	298.0637236	3.135806382	219.7880541
2086/01/01	0.666624504	7.1558332	329.8813376	295.8950554	3.184788320	282.6062251
2087/01/01	0.665366386	7.1344985	329.9160895	295.8499387	3.181153526	345.8284325
2088/01/01	0.665453128	7.1342415	329.9271610	295.8394110	3.181596889	49.2633623
2088/12/31	0.665699992	7.1373335	329.9675060	295.7944097	3.181734181	112.6624583
2090/01/01	0.666061244	7.1451269	330.0293875	295.7418704	3.181353761	176.0327377
2091/01/01	0.666281456	7.1517650	330.0785801	295.7186856	3.180600695	239.4035326
2092/01/01	0.666295044	7.1537441	330.1096290	295.7164796	3.179339403	302.8160418
2092/12/31	0.666234418	7.1534072	330.1287226	295.7141615	3.178359319	6.3170365
2094/01/01	0.666433394	7.1537144	330.1172525	295.6574905	3.176259088	69.9175193
2095/01/01	0.667064308	7.1647708	330.1357158	295.5443138	3.172746982	133.7306378
2096/01/01	0.669570031	7.2206290	330.3231148	295.2709716	3.167490261	197.7651225
2096/12/31	0.680500900	7.5395336	331.6221077	294.6493073	3.135954725	261.1071415
2098/01/01	0.680856184	7.5546267	332.0892619	294.6226642	3.115441269	325.4128758
2099/01/01	0.680864089	7.5535356	332.1144600	294.6132523	3.114969800	30.8825431
2100/01/01	0.681043486	7.5548197	332.1191023	294.5907174	3.114588416	96.3689035
2101/01/01	0.681376986	7.5592262	332.1418360	294.5610330	3.114052195	161.8532628

141P/Machholz 2

Date	e	i	w	Node	a	M0
2010/01/01	0.748925788	12.8029570	149.3688172	246.0901708	3.017953007	332.9549972
2011/01/01	0.748995055	12.8023676	149.3827697	246.0778158	3.019493164	41.5857379
2012/01/01	0.748988366	12.8072682	149.4397741	246.0433939	3.020755071	110.1133462
2012/12/31	0.748772802	12.8137201	149.4862374	246.0196405	3.021540205	178.5852095
2014/01/01	0.748471386	12.8149508	149.4958353	246.0164357	3.021838977	247.0644287
2015/01/01	0.748252177	12.8104539	149.4811667	246.0178542	3.022165580	315.5815651
2016/01/01	0.748273130	12.8088862	149.4882400	246.0090132	3.023123005	24.0883060
2016/12/31	0.748998244	12.8813964	150.0550778	245.4642705	3.027340542	92.2894599
2018/01/01	0.745782208	13.7175252	153.2065472	242.4616557	3.038546200	159.8180021
2019/01/01	0.738318198	13.9845248	153.7064747	241.8813939	3.052128596	227.6420121
2020/01/01	0.736202155	13.9496157	153.5486314	241.8931259	3.057689732	295.1699351
2020/12/31	0.735776535	13.9335547	153.5283797	241.8680144	3.057875560	2.4797157
2022/01/01	0.735947811	13.9349642	153.5471238	241.8467585	3.058897156	69.7600514
2023/01/01	0.736150482	13.9429079	153.5821309	241.8136978	3.059056681	136.9996178
2024/01/01	0.736199811	13.9505820	153.6156524	241.7957691	3.059051974	204.2083141
2024/12/31	0.736091529	13.9522419	153.6293196	241.7936787	3.058722531	271.4362270
2026/01/01	0.735945150	13.9492510	153.6314942	241.7907651	3.058192394	338.7130508
2027/01/01	0.736073902	13.9495992	153.6460355	241.7608140	3.058444044	46.0166428
2028/01/01	0.736607522	13.9708299	153.6857744	241.6476498	3.055763204	113.4471554
2028/12/31	0.738064636	14.0484639	153.8247490	241.4448368	3.052541891	181.0279278
2030/01/01	0.742744975	14.3845596	154.3874199	241.0444162	3.042133532	248.4538718
2031/01/01	0.744147833	14.5372293	154.8177923	240.9865134	3.024360182	315.8175512
2032/01/01	0.744078777	14.5352691	154.8463424	240.9809913	3.022566603	24.2795938
2032/12/31	0.744220364	14.5381806	154.8416141	240.9627384	3.021728785	92.8156783
2034/01/01	0.744524677	14.5473657	154.8600514	240.9354791	3.021209969	161.3493316
2035/01/01	0.744776498	14.5588154	154.8914678	240.9170857	3.020630747	229.8592164
2036/01/01	0.744842469	14.5643823	154.9147774	240.9149512	3.019674706	298.3905084
2036/12/31	0.744885696	14.5632640	154.9278772	240.9125645	3.019828242	6.9912413
2038/01/01	0.744803710	14.5627345	154.9271107	240.9033874	3.018870300	75.6326574
2039/01/01	0.744703297	14.5626704	154.9164447	240.9040408	3.018461586	144.3276343
2040/01/01	0.744785142	14.5665790	154.9058906	240.8975144	3.018312944	213.0570734
2040/12/31	0.745238907	14.5813366	154.9190414	240.8894756	3.018054120	281.7751415
2042/01/01	0.745473755	14.5876008	154.9410596	240.8943551	3.017522005	350.4469964
2043/01/01	0.745422895	14.5878491	154.9352362	240.8926999	3.016968173	59.1447151
2044/01/01	0.745439011	14.5899519	154.9262461	240.8852153	3.016147463	127.9070994
2044/12/31	0.745656041	14.5975090	154.9324109	240.8692562	3.015655876	196.6738310
2046/01/01	0.745944215	14.6087059	154.9547442	240.8587440	3.015015277	265.4163557
2047/01/01	0.746080554	14.6142335	154.9752834	240.8617660	3.014149335	334.1754856
2048/01/01	0.746090896	14.6144203	154.9838352	240.8597839	3.014777936	42.9566281
2048/12/31	0.745893486	14.6139082	155.0049982	240.8617018	3.015600524	111.6935523
2050/01/01	0.745544032	14.6078748	154.9978739	240.8767820	3.016233955	180.4274641
2051/01/01	0.745271836	14.6001521	154.9753606	240.8864540	3.016882871	249.1713184
2052/01/01	0.745216466	14.5959728	154.9550488	240.8866494	3.017894886	317.8979130
2052/12/31	0.745221363	14.5953706	154.9545426	240.8815536	3.018505113	26.5561094
2054/01/01	0.744736966	14.6023043	155.0559539	240.8374668	3.020462857	95.1400415
2055/01/01	0.744008683	14.6037748	155.0692267	240.8314548	3.021277400	163.7805297
2056/01/01	0.743855244	14.6018297	155.0437411	240.8348400	3.021757422	232.4204576

116

Date	e	i	w	Node	a	M0
2056/12/31	0.743966346	14.6029941	155.0311046	240.8348595	3.022450340	301.0061382
2058/01/01	0.744024362	14.6007367	155.0134727	240.8265578	3.023669416	9.5150022
2059/01/01	0.744060628	14.6024646	155.0537496	240.8063296	3.025066437	77.9242855
2060/01/01	0.743941990	14.6060738	155.0888353	240.7933469	3.025952636	146.2657473
2060/12/31	0.743627173	14.6033777	155.0960901	240.7976511	3.026518695	214.5938243
2062/01/01	0.743329378	14.5964793	155.0834220	240.8016449	3.026904555	282.9415262
2063/01/01	0.743211975	14.5918985	155.0719270	240.7957927	3.027350843	351.2962063
2064/01/01	0.743540062	14.6022446	155.2032732	240.6913887	3.031277227	59.5092642
2064/12/31	0.743881355	14.9952289	156.6827989	239.2918661	3.037567038	127.1128198
2066/01/01	0.735022393	15.7034329	158.2867706	237.7981946	3.054981337	194.5341594
2067/01/01	0.731124089	15.6830898	158.1375403	237.7995663	3.063211317	261.9701393
2068/01/01	0.730151830	15.6427504	158.0611004	237.7706486	3.066233174	329.1039230
2068/12/31	0.730149764	15.6403225	158.0757298	237.7543287	3.066584972	36.1427863
2070/01/01	0.730341219	15.6480409	158.1121935	237.7174069	3.067108091	103.1293416
2071/01/01	0.730500018	15.6623866	158.1599041	237.6840122	3.067291345	170.0633500
2072/01/01	0.730466728	15.6709848	158.1908145	237.6735908	3.067055052	236.9863325
2072/12/31	0.730321133	15.6696384	158.2007618	237.6727044	3.066462475	303.9618259
2074/01/01	0.730245506	15.6677857	158.1994770	237.6642889	3.066032072	11.0018537
2075/01/01	0.730540775	15.6777882	158.2264444	237.5860303	3.064723591	78.0948665
2076/01/01	0.731462897	15.7290078	158.2907764	237.4418641	3.061681866	145.3636398
2076/12/31	0.734431159	15.9307471	158.5352774	237.1725084	3.056262736	212.7500119
2078/01/01	0.739652603	16.3616945	159.2474050	236.9832831	3.033180358	279.3594844
2079/01/01	0.739346216	16.3505426	159.3980943	236.9673069	3.024252521	347.3847764
2080/01/01	0.739430803	16.3510902	159.3944567	236.9588996	3.024359806	55.8292912
2080/12/31	0.739712971	16.3593264	159.4000062	236.9322845	3.023658144	124.2919633
2082/01/01	0.740049067	16.3727817	159.4244641	236.9097729	3.023125066	192.7371125
2083/01/01	0.740241883	16.3834193	159.4522581	236.9027114	3.022328179	261.1737030
2084/01/01	0.740236264	16.3853761	159.4725721	236.9035452	3.021101843	329.6619635
2084/12/31	0.740237852	16.3848962	159.4713819	236.8995486	3.020812024	38.2240327
2086/01/01	0.740116183	16.3843964	159.4634073	236.9009600	3.020159904	106.8415730
2087/01/01	0.740049388	16.3836816	159.4472200	236.9031748	3.019915601	175.5036097
2088/01/01	0.740290814	16.3927030	159.4407780	236.8966907	3.019806005	244.1826685
2088/12/31	0.740784076	16.4085383	159.4681863	236.9014009	3.018956192	312.8121638
2090/01/01	0.740851962	16.4092873	159.4734036	236.9024067	3.018893631	21.4439911
2091/01/01	0.740795546	16.4096988	159.4622858	236.9004793	3.017893607	90.1279843
2092/01/01	0.740977636	16.4146244	159.4529864	236.8900248	3.017094879	158.8602056
2092/12/31	0.741302006	16.4252748	159.4632042	236.8784604	3.016522830	227.5798611
2094/01/01	0.741570859	16.4357726	159.4850116	236.8780554	3.015692652	296.2865251
2095/01/01	0.741685887	16.4375290	159.4920658	236.8804778	3.016019269	5.0286220
2096/01/01	0.741605128	16.4374501	159.5147007	236.8764636	3.016693118	73.7366381
2096/12/31	0.741352901	16.4349735	159.5242142	236.8827904	3.017308803	142.4206272
2098/01/01	0.741053983	16.4273295	159.5101026	236.8933617	3.017839385	211.1193533
2099/01/01	0.740915066	16.4214912	159.4907615	236.8955309	3.018539020	279.8184904
2100/01/01	0.740957227	16.4198964	159.4777419	236.8936425	3.019640139	348.4734111
2101/01/01	0.740861930	16.4212874	159.5103365	236.8814314	3.020516151	57.0588452

COMETE NON NUMERATE E
POSSIBILI RITORNI
UNNUMBERED COMETS

```
Comet                        (Possible) data of return
D/1766 G1 (Helfenzrieder)        2013
D/1770 L1 (Lexell)               2184
D/1819 W1 (Blanpain)             2013
D/1884 O1 (Barnard)              2015
D/1886 K1 (Brooks)               2014
D/1894 F1 (Denning)              2015
D/1895 Q1 (Swift)                2019
C/1917 F1 (Mellish)              2061
D/1918 W1 (Schorr)               2018
C/1921 H1 (Dubiago)
C/1937 D1 (Wilk)                 2121
C/1942 EA (Vaisala)
D/1952 B1 (Harrington-Wilson)    2017
D/1977 C1 (Skiff-Kosai)          2014
D/1978 R1 (Haneda-Campos)        2016
D/1993 F2 (Shoemaker-Levy)
C/1984 A1 (Bradfield)            2126
C/1989 A3 (Bradfield)
C/1991 L3 (Levy)                 2042
P/1994 N2 (McNaught-Hartley)     2015
P/1996 R2 (Lagerkvist)           2011
P/1997 B1 (Kobayashi)            2022
P/1997 G1 (Montani)              2016
P/1997 T3 (Lagerkvist-Carsenty)  2015
C/1998 G1 (LINEAR)               2040
P/1998 QP54 (LONEOS-Tucker)      2015
P/1998 U3 (Jager)                2014
P/1998 VS24 (LINEAR)             2018
C/1998 Y1 (LINEAR)               2109
P/1998 Y2 (Li)                   2014
C/1999 E1 (Li)                   2065
C/1999 G1 (LINEAR)               2131
P/1999 RO28 (LONEOS)             2012
C/1999 S3 (LINEAR)               2080
P/1999 V1 (Catalina)             2016
P/1999 XN120 (Catalina)          2017
C/1999 XS87 (LINEAR)             2070
C/2000 D2 (LINEAR)               2072
C/2000 G2 (LINEAR)               2052
P/2000 QJ46 (LINEAR)             2014
P/2000 R2 (LINEAR)               2013
P/2000 S1 (Skiff)                2017
C/2000 S3 (LONEOS)               2040
P/2000 S4 (LINEAR-Spacewatch)    2019
P/2001 BB50 (LINEAR-NEAT)        2014
P/2001 F1 (NEAT)                 2017
P/2001 H5 (NEAT)                 2015
C/2001 M10 (NEAT)                2138
C/2001 OG108 (LONEOS)            2050
P/2001 Q6 (NEAT)                 2024
P/2001 Q11 (NEAT)                2014
P/2001 R6 (LINEAR-Skiff)         2010
P/2001 T3 (NEAT)                 2018
C/2001 W2 (BATTERS)              2076
```

```
Comet                   (Possible) data of return
C/2002 A1 (LINEAR)           2066
C/2002 A2 (LINEAR)           2065
P/2002 AR2 (LINEAR)          2014
C/2002 B1 (LINEAR)           2033
C/2002 CE10 (LINEAR)         2033
P/2002 EJ57 (LINEAR)         2018
C/2002 K4 (NEAT)             2075
P/2002 Q1 (Van Ness)         2015
P/2002 T5 (LINEAR)           2021
P/2002 T6 (NEAT-LINEAR)      2024
C/2003 E1 (NEAT)             2055
C/2003 F1 (LINEAR)           2096
P/2003 F2 (NEAT)             2019
P/2003 L1 (Scotti)           2020
P/2003 O3 (LINEAR)           2014
P/2003 QX29 (NEAT)           2025
C/2003 R1 (LINEAR)           2089
P/2003 S1 (NEAT)             2013
P/2003 S2 (NEAT)             2011
P/2003 SQ215 (NEAT-LONEOS)   2017
P/2003 T12 (SOHO)            2016
C/2003 U1 (LINEAR)           2110
P/2003 U2 (LINEAR)           2013
P/2003 U3 (NEAT)             2014
C/2003 W1 (LINEAR)           2128
P/2003 WC7 (LINEAR-Catalina) 2027
P/2004 A1 (LONEOS)           2027
C/2004 C1 (Larsen)           2043
P/2004 DO29 (Spacewatch-LINEAR) 2024
P/2004 FY140 (LINEAR)        2015
P/2004 R1 (McNaught)         2010
P/2004 R3 (LINEAR-NEAT)      2011
P/2004 T1 (LINEAR-NEAT)      2011
P/2004 V1 (Skiff)            2014
P/2004 V3 (Siding Spring)    2023
P/2004 V5 (LINEAR-Hill)      2027
P/2004 VR8 (LONEOS)          2016
P/2004 WR9 (LINEAR)          2020
P/2005 E1 (Tubbiolo)         2023
P/2005 GF8 (LONEOS)          2019
P/2005 J1 (McNaught)         2012
P/2005 JD108 (Catalina-NEAT) 2021
P/2005 JN (Spacewatch)       2012
P/2005 JQ5 (Catalina)        2014
P/2005 L1 (McNaught)         2013
P/2005 L4 (Christensen)      2014
C/2005 N5 (Catalina)         2160
C/2005 O2 (Christensen)      2117
P/2005 Q4 (LINEAR)           2015
P/2005 R1 (NEAT)             2018
P/2005 RV25 (LONEOS-Christensen) 2015
P/2005 S2 (Skiff)            2029
P/2005 S3 (Read)             2016
P/2005 SB216 (LONEOS)        2026
```

```
Comet                    (Possible) data of return
P/2005 T2 (Christensen)        2012
P/2005 T3 (Read)               2026
P/2005 T4 (SWAN)               2034
P/2005 T5 (Broughton)          2025
C/2005 W2 (Christensen)        2087
P/2005 W3 (Kowalski)           2021
P/2005 XA54 (LONEOS-Hill)      2021
P/2005 Y2 (McNaught)           2020
P/2006 D1 (Hill)               2018
P/2006 F1 (Kowalski)           2018
C/2006 F2 (Christensen)        2049
P/2006 F4 (Spacewatch)         2012
P/2006 G1 (McNaught)           2017
P/2006 H1 (McNaught)           2019
P/2006 HR30 (Siding Spring)    2028
P/2006 R1 (Siding Spring)      2019
P/2006 R2 (Christensen)        2014
P/2006 S1 (Christensen)        2013
P/2006 S4 (Christensen)        2022
P/2006 S6 (Hill)               2015
C/2006 U7 (Gibbs)              2049
P/2006 W1 (Gibbs)              2020
P/2006 XG16 (Spacewatch)       2014
P/2007 A2 (Christensen)        2022
P/2007 B1 (Christensen)        2021
P/2007 C1 (Christensen)        2013
P/2007 C2 (Catalina)           2026
P/2007 H1 (McNaught)           2014
P/2007 H3 (Garradd)            2014
P/2007 K2 (Gibbs)              2026
P/2007 Q2 (Gilmore)            2020
P/2007 R1 (Larson)             2027
P/2007 R2 (Gibbs)              2014
P/2007 R3 (Gibbs)              2016
P/2007 R4 (Garradd)            2021
P/2007 R5 (SOHO)               2011
P/2007 S1 (Zhao)               2015
C/2007 S2 (Lemmon)             2052
P/2007 T2 (Kowalski)           2013
P/2007 T4 (Gibbs)              2019
P/2007 T6 (Catalina)           2017
P/2007 V1 (Larson)             2019
P/2007 V2 (Hill)               2015
P/2007 VQ11 (CATALINA)         2020
P/2008 A2 (LINEAR)             2014
P/2008 CL94 (Lemmon)           2021
C/2008 E1 (Catalina)           2043
P/2008 J2 (Beshore)            2014
P/2008 J3 (McNaught)           2016
P/2008 L2 (Hill)               2023
P/2008 O2 (McNaught)           2018
P/2008 O3 (Boattini)           2031
P/2008 Q2 (Ory)                2014
P/2008 QP20 (LINEAR-Hill)      2015
```

```
Comet                      (Possible) data of return
C/2008 R3 (LINEAR)                    2087
P/2008 S1 (Catalina-McNaught)         2015
C/2008 S2 (SOHO)                      2012
P/2008 T1 (Boattini)                  2016
P/2008 T4 (Hill)                      2018
P/2008 WZ96 (LINEAR)                  2015
P/2008 Y1 (Boattini)                  2019
P/2008 Y2 (Gibbs)                     2015
P/2008 Y3 (McNaught)                  2031
P/2009 B1 (Boattini)                  2026
P/2009 K1 (Gibbs)                     2016
C/2009 K4 (Gibbs)                     2279
P/2009 L2 (Yang-Gao)                  2015
P/2009 O3 (Hill)                      2031
P/2009 Q1 (Hill)                      2022
P/2009 Q4 (Boattini)                  2015
P/2009 Q5 (McNaught)                  2029
P/2009 S2 (McNaught)                  2017
P/2009 SK280 (Spacewatch-Hill)        2019
P/2009 T2 (La Sagra)                  2010
P/2009 U4 (McNaught)                  2021
P/2009 WX51 (Catalina)                2010
P/2009 Y2 (Kowalski)                  2010
P/2010 A1 (Hill)                      2018
P/2010 A2 (LINEAR)                    2013
P/2010 A3 (Hill)                      2010
P/2010 A5 (LINEAR)                    2010
P/2010 B2 (WISE)                      2015
P/2010 C1 (Scotti)                    2028
P/2010 D1 (WISE)                      2017
P/2010 D2 (WISE)                      2010
P/2010 E2 (Jarnac)                    2010
P/2010 H5 (Scotti)                    2010
P/2010 J3 (McMillan)                  2010
P/2010 J5 (McNaught)                  2018
P/2010 JC81 (WISE)                    2011
P/2010 N1 (WISE)                      2010
P/2010 P4 (WISE)                      2010
P/2010 R2 (La Sagra)                  2010
P/2010 T2 (PANSTARRS)                 2011
P/2010 TO20 (LINEAR-Grauer)           2022
P/2010 U1 (Boattini)                  2010
P/2010 U2 (Hill)                      2010
P/2010 UH55 (Spacewatch)              2011
P/2010 V1 (Ikeya-Murakami)            2010
P/2010 WK (LINEAR)                    2010
P/2011 A2 (Scotti)                    2010
P/2011 C2 (Gibbs)                     2012
P/2011 FR143 (Lemmon)                 2011
C/2011 J3 (LINEAR)                    2011
P/2011 JB15 (Spacewatch-Boattini)2012
C/2011 L1 (McNaught)                  2010
P/2011 N1 (ASH)                       2012
P/2011 P1 (McNaught)                  2010
```

```
Comet                    (Possible) data of return
P/2011 Q3 (McNaught)              2011
P/2011 R3 (Novichonok-Gerke)     2012
P/2011 S1 (Gibbs)                2014
C/2011 S2 (Kowalski)             2011
P/2011 U1 (PANSTARRS)            2012
P/2011 U2 (Bressi)               2012
P/2011 UA134 (Spacewatch-PANSTARRS)2011
P/2011 VJ5 (Lemmon)              2011
P/2011 W1 (PANSTARRS)            2012
P/2011 W2 (Rinner)               2011
P/2011 Y2 (Boattini)             2012
C/2011 Y3 (Boattini)             2011
P/2012 B1 (PANSTARRS)            2013
C/2012 BJ98 (Lemmon)             2012
P/2012 F2 (PANSTARRS)            2013
C/2012 H2 (McNaught)             2012
P/2012 K3 (Gibbs)                2012
P/2012 NJ (La Sagra)             2012
P/2012 O1 (McNaught)             2012
P/2012 O2 (McNaught)             2012
P/2012 O3 (McNaught)             2012
C/2012 Q1 (Kowalski)             2012
P/2012 S2 (La Sagra)             2012
P/2012 SB6 (Lemmon)              2012
P/2012 T1 (PANSTARRS)            2012
C/2012 T6 (Kowalski)             2012
C/2012 US27 (Siding Spring)      2013
P/2012 WA34 (Lemmon-PANSTARRS)   2013
C/2012 X2 (PANSTARRS)            2013
C/2012 Y3 (McNaught)             2012
P/2013 A2 (Scotti)               2013
C/2013 D1 (Holvorcem)            2013
P/2013 EV9 (Spacewatch)          2013
```

GRANDI COMETE DEL PASSATO
GREATS COMETS IN THE PAST

```
1st Date      Obs     Perihelion          Perigee        Brightness Max.
Reported      Int    Date      Dist     Date      Dist    Date       Mag  Name, Notes
YYYY/mmm/DD   (d)    YYYY/mmm/DD (AU)    YYYY/mmm/DD (AU)  YYYY/mmm/DD
------------  ---    ------------ -----  ----------- ----  ----------- ----  ----------------------
Julian Calendar

B.C. dates
373-372 Winter                                                          1
    87/Jul      35    87/Aug/06 0.59     87/Jul/27  0.44   87/Jul/27   2  1P/Halley
    12/Aug/25   57    12/Oct/10 0.59     12/Sep/10  0.16   12/Sep/10   1  1P/Halley

A.D. Dates
    66/Jan/30   71    66/Jan/26 0.59     66/Mar/20  0.25   66/Mar/20   1  1P/Halley
   141/Mar/26   41   141/Mar/22 0.58    141/Apr/22  0.17  141/Apr/22  -1  1P/Halley
   178/Sep      80                                                       2
   191/Oct                                                               2
   218/May      40   218/May/17 0.58    218/May/30  0.42  218/May/30   0  1P/Halley
   240/Nov/10   39   240/Nov/10 0.37    240/Nov/30  1.00  240/Nov/20  1-2  (240 V1)
   295/May      30   295/Apr/20 0.58    295/May/12  0.32  295/May/12   0  1P/Halley
   374/Mar/01   32   374/Feb/16 0.58    374/Apr/02  0.09  374/Apr/02  -1  1P/Halley
   390/Aug/21   26   390/Sep/05 0.92    390/Aug/18  0.10  390/Aug/18  -1  (390 Q1), 2
   400/Mar/18   30   400/Feb/25 0.21    400/Mar/31  0.08  400/Mar/19   0  (400 F1)
   442/Nov/09  100   442/Dec/15 1.53    442/Dec/07  0.58  442/Dec/07  1-2  (442 V1)
   451/Jun/09   68   451/Jun/28 0.58    451/Jun/30  0.49  451/Jun/30   0  1P/Halley
   565/Jul/22  100   565/Jul/15 0.82    565/Sep/15  0.54  565/Sep/13  0-1  (565 O1)
   568/Jul/28  106   568/Aug/27 0.87    568/Sep/25  0.09  568/Sep/25   0  (568 O1)
   607/Mar-Apr  30   607/Mar/15 0.58    607/Apr/19  0.09  607/Apr/19  -2  1P/Halley
   684/Sep/06   33   684/Oct/02 0.58    684/Sep/07  0.26  684/Sep/07  1-2  1P/Halley
   760/May/16   50   760/May/20 0.58    760/Jun/03  0.41  760/Jun/03   0  1P/Halley
   770/May/25   62   770/Jun/05 0.58    770/Jul/10  0.30  770/Jul/10  1-2  (770 K1)
   837/Mar/21   39   837/Feb/28 0.58    837/Apr/11  0.03  837/Apr/11  -3  1P/Halley, 3
   838/Nov/09   49
   891/May/12   62                                                       2
   905/May/18   26   905/Apr/26 0.20    905/May/25  0.21  905/May/23   0  2
   989/Aug/10   32   989/Sep/05 0.58    989/Aug/20  0.39  989/Aug/20  1-2  1P/Halley
  1066/Apr/02   66  1066/Mar/20 0.58   1066/Apr/24  0.10 1066/Apr/24  -1  1P/Halley
  1106/Feb/02   40                                                       4
  1132/Oct/03   24  1132/Aug/30 0.74   1132/Oct/07  0.04 1132/Oct/07  -1  (1132 T1)
  1145/Apr/15   65  1145/Apr/18 0.58   1145/May/12  0.27 1145/May/12   0  1P/Halley
  1222/Sep/02   36  1222/Sep/28 0.58   1222/Sep/06  0.31 1222/Sep/24  1-2  1P/Halley, 5
  1240/Jan/27   64  1240/Jan/21 0.67   1240/Feb/02  0.36 1240/Feb/02   0  (1240 B1)
  1264/Jul/17   85  1264/Jul/20 0.82   1264/Jul/29  0.18 1264/Jul/29   0  (1264 N1), 6
  1301/Sep/01   61  1301/Oct/25 0.58   1301/Sep/23  0.18 1301/Sep/23  1-2  1P/Halley
  1378/Sep/26   15  1378/Nov/10 0.58   1378/Oct/03  0.12 1378/Oct/03   1  1P/Halley, 7
  1402/Feb/08   70  1402/Mar/21 0.38   1402/Feb/19  0.71 1402/Mar/12  -3  (1402 D1), 8
  1456/May/26   44  1456/Jun/09 0.58   1456/Jun/19  0.45 1456/Jun/19   0  1P/Halley
  1468/Sep/18   56  1468/Oct/07 0.85   1468/Oct/02  0.67 1468/Oct/02  1-2  (1468 S1)
  1471/Dec/25   23  1472/Mar/01 0.49   1472/Jan/23  0.49 1472/Jan/23  -3  (1471 Y1)
  1531/Aug/05   34  1531/Aug/26 0.58   1531/Aug/14  0.44 1531/Aug/27   1  1P/Halley
  1532/Sep/02  120  1532/Oct/18 0.52   1532/Sep/21  0.67 1532/Oct/13  -1  (1532 R1)
  1533/Jun/27   82  1533/Jun/15 0.25   1533/Aug/02  0.42 1533/Jun/27   0  (1533 M1), 9
  1556/Feb/27   72  1556/Apr/22 0.49   1556/Mar/13  0.08 1556/Mar/14  -2  (1556 D1)
  1577/Nov/01   87  1577/Oct/27 0.18   1577/Nov/10  0.63 1577/Nov/08  -3  (1577 V1)
  1618/Nov/16   67  1618/Nov/08 0.39   1618/Dec/06  0.36 1618/Nov/29  0-1  (1618 W1)
  1664/Nov/17   75  1664/Dec/04 1.03   1664/Dec/29  0.17 1664/Dec/29  -1  (1664 W1)
  1665/Mar/27   24  1665/Apr/24 0.11   1665/Apr/04  0.57 1665/Apr/20  -1  (1665 F1), 10
  1668/Mar/03   27  1668/Feb/28 0.07   1668/Mar/05  0.80 1668/Mar/08  1-2  (1668 E1)
  1680/Nov/23   88  1680/Dec/18 0.01   1680/Nov/30  0.42 1680/Dec/29  1-2  (1680 V1), 11
  1682/Aug/15   41  1682/Sep/15 0.58   1682/Aug/31  0.42 1682/Aug/31  0-1  1P/Halley
  1686/Aug/12   34  1686/Sep/16 0.34   1686/Aug/16  0.32 1686/Aug/31  1-2  (1686 R1)
  1743/Nov/29  110  1744/Mar/01 0.22   1744/Feb/27  0.83 1744/Feb/20  -3  (1743 X1), 12
  1769/Aug/24   94  1769/Oct/08 0.12   1769/Sep/10  0.32 1769/Sep/22   0  Messier (1769 P1), 13
  1807/Sep/09   90  1807/Sep/19 0.65   1807/Sep/27  1.15 1807/Sep/20   2  Great Comet (1807 R1)
  1811/Apr/11  260  1811/Sep/12 1.04   1811/Oct/16  1.22 1811/Oct/20   0  Great Comet (1811 F1)
  1843/Feb/05   48  1843/Feb/27 0.006  1843/Mar/06  0.84 1843/Mar/07  <-3  Great March Comet (1843 D1),
14
  1858/Aug/20   80  1858/Sep/30 0.58   1858/Oct/11  0.54 1858/Oct/07  0-1  Donati (1858 L1)
  1861/May/13   90  1861/Jun/12 0.82   1861/Jun/30  0.13 1861/Jun/27   0  Great Comet (1861 J1), 13
  1865/Jan/17   36  1865/Jan/14 0.03   1865/Jan/16  0.94 1865/Jan/24   1  Great Southern Comet (1865
B1), 15
  1874/Jun/10   50  1874/Jul/09 0.68   1874/Jul/23  0.29 1874/Jul/13  0-1  Coggia (1874 H1)
  1882/Sep/01  135  1882/Sep/17 0.008  1882/Sep/16  0.99 1882/Sep/08  <-3  Great September Comet (1882
R1), 16
  1901/Apr/12   38  1901/Apr/24 0.24   1901/Apr/30  0.83 1901/May/05   1  Great Comet (1901 G1)
  1910/Jan/13   20  1910/Jan/17 0.13   1910/Jan/18  0.86 1910/Jan/30  1-2  Great January Comet (1910
A1), 17
  1910/Apr/10   80  1910/Apr/20 0.59   1910/May/20  0.15 1910/May/20  0-1  1P/Halley
  1927/Nov/27   32  1927/Dec/18 0.18   1927/Dec/12  0.75 1927/Dec/08   1  Skjellerup-Maristany (1927
X1), 18
  1965/Oct/03   30  1965/Oct/21 0.008  1965/Oct/17  0.91 1965/Oct/14   2  Ikeya-Seki (1965 S1), 19
  1970/Feb/10   80  1970/Mar/20 0.54   1970/Mar/26  0.69 1970/Mar/20  0-1  Bennett (1969 Y1), 20
  1976/Feb/05   55  1976/Feb/25 0.20   1976/Feb/29  0.79 1976/Mar/01  -1  West (1975 V1), 21
  1996/Mar/15   30  1996/May/01 0.23   1996/Mar/25  0.10 1996/Apr/20  1-2  Hyakutake (1996 B2)
  1996/Sep/09  215  1997/Apr/01 0.91   1997/Mar/22  1.32 1997/Mar/26  -0.7  Hale-Bopp (1995 O1), 22
  2007/Jan/01   25  2007/Jan/12 0.17   2007/Jan/15  0.82 2007/Jan/14  -6  McNaught (2006 P1), 23
```

125

Notes

1. Reported by the Greek historian Ephorus to have split into two pieces.
2. The Chinese reported that the tail spanned more than 70 degrees.
3. The closest approach to the Earth that comet Halley has ever made. On Apr. 13, the comet's tail was more than 90 degrees in
length.
4. This comet passed very close to the sun and is perhaps the progenitor of the sungrazing comets of 1882 and 1965 or that of 1843.
5. Korean observers reported the comet was visible during the daylight hours on September 9th (probably during twilight only).
6. On July 26, Chinese observers reported the tail spanning 100 degrees.
7. Chinese observers reported cloudy weather from October 11 until Nov. 9, at which time the comet had passed behind the sun.
8. In mid-March, the comet entered solar conjunction and there were reports that it was a daylight object for 8 days.
9. The comet was discovered emerging from solar conjunction.
10. Last observed on April 20 as it approached solar conjunction.
11. This was the first comet discovered with the aid of a telescope (on Nov. 14).
12. Visible in daylight only 12 degrees from the Sun on February 27.
13. Tail reported as longer than 90 degrees near Earth close approach.
14. On the date of perihelion, this sungrazing comet was observed in daylight nearly one degree from the sun.
15. Comet observed in southern hemisphere.
16. The Great September comet was a brilliant object that was observed very close to the sun, and split into at least four separate
pieces near perihelion. This comet and comet Ikeya-Seki in 1965 are believed to be members of the same family of sungrazing
comets.
17. This comet was easily observed on January 17 only 4.5 degrees from the sun. It is often confused with the later apparition of
comet Halley in mid-1910.

PASSAGGI RAVVICINATI DI COMETE
NEAR EARTH COMETS DATES

Object (and name)	Date of encounter (TT)			Distance
	JD	Calendar		(AU)
209P/LINEAR	2456806.83	2014 May	29.33	0.05545
252P/LINEAR	2457469.02	2016 Mar.	21.52	0.03575
45P/Honda-Mrkos-Pajdusak	2457795.76	2017 Feb.	11.26	0.08426
46P/Wirtanen	2458469.19	2018 Dec.	16.69	0.07789
P/2006 U1 (LINEAR)	2462443.49	2029 Nov.	2.99	0.04624
249P/LINEAR	2462444.48	2029 Nov.	3.98	0.05704
252P/LINEAR	2463307.29	2032 Mar.	15.79	0.05847
P/2005 JQ5 (Catalina)	2464854.07	2036 June	9.57	0.04663
P/2009 WX51 (Catalina)	2465166.50	2037 Apr.	18.00	0.05224

INDICE - INDEX

www.ingramcontent.com/pod-product-compliance
Lightning Source LLC
Chambersburg PA
CBHW022006170526
45157CB00003B/1169